Five Emus to the Kin

Environment and E

Cross Cultures

Readings in the Post / Colonial Literatures in English

92

Five Emus to the King of Siam

Environment and Empire

Edited by
Helen Tiffin

Rodopi

Amsterdam - New York, NY 2007

Cover design: Gordon Collier and Pier Post
Photo's by Helen Tiffin

The paper on which this book is printed meets the requirements of "ISO 9706:1994, Information and documentation - Paper for documents - Requirements for permanence".

ISBN: 978-90-420-2243-0

Printed in The Netherlands

Table of Contents

Acknowledgements

Great thanks are due to Simone Murray, without whose careful and intelligent editing this volume would never have been completed. I am also particularly grateful to Gordon Collier for his editing and his unstinting patience under extreme provocation. To English Department colleagues at the University of Queensland, especially former members of the Post-Colonial Studies Research Group, my warmest thanks and continuing appreciation for intellectual stimulation, interest and belief in the importance of issues above and beyond the "mere professionalism" increasingly affecting the academy. Above all, my most profound thanks to my former postgraduate students, all of whom have, over the years, made my life personally, intellectually and politically so much richer and brighter.

Illustrations

Introduction

HELEN TIFFIN

As LAWRENCE BUELL has aptly remarked, "criticism worthy of its name arises from commitments deeper than professionalism,"[1] and in the genesis and subsequent practice of the relatively new fields of postcolonial and environmental studies, such commitment has usually been both evident and energizing. Postcolonial critiques of European imperialism and colonialism (and studies of their post-independence legacies) have from the outset been informed by ethical and political concerns; while the burgeoning area of environmental analysis and critique, particularly in the humanities, has in large part emerged out of genuine alarm at the future of the planetary environment and its inhabitants. Such concerns come in the wake of a taken-for-granted human domination where anthropocentrism and Western imperialism are intrinsically interwoven. Consequently, both postcolonial critique and environmental criticism have been, and remain, again in Buell's words, "deeply polemical"[2] while maintaining their commitment to a rigorous scholarship; one which, however, like scientific 'objectivity', is increasingly compromised by instrumentalist pressures on the Academy.

Using Ralph Waldo Emerson's writing on slavery as his example, Buell compares the urgency of Emerson's address to "an issue widely recognized as grave, yet not so widely believed to require immediate action" with the

---

[1] Lawrence Buell, *The Future of Environmental Criticisms* (Oxford: Blackwell, 2005): 97.

[2] Buell, *The Future of Environmental Criticisms*, 97.

position of contemporary environmental scholars. Emerson knew, so Buell argues, that his readers would agree that slavery was "the greatest public evil in the nation's history [... but ...] only a few were as yet prepared to do anything about it."[3] (97). The comparison between this philosophical and polemical engagement with the wrongs of slavery and contemporary concern over the destruction and/or degradation of environments and non-human animals is an instructive one. Both pose (or posed) arguments against an entrenched (and highly commercialized) status quo; but just as significantly, human slavery and environmental damage are connected because human – and, more specifically, Western – exploitation of other peoples is inseparable from attitudes and practices in relation to *other species* and the extra-human environment generally.

Over three-quarters of the world has been or continues to be radically affected by the era of European imperialism and colonization. This history has generally been read and understood in primarily human terms, whether militarily, politically, socially, philosophically or culturally. In fact, however, both conquest and colonization were frequently facilitated by, and in turn radically affected, the colonized through extra-human avenues, as Alfred Crosby has argued in his groundbreaking *The Columbian Exchange* (1972) and *Ecological Imperialism* (1986). Writing from a rather different perspective, the historian Richard Grove also notes that "in many ways the business of empire, for most of the colonized, had far more to do with the impact of different modes of colonial resource control and colonial environmental concepts, than it had to do with the direct impact of military or political structures."[4] Imported diseases, destruction of wildlife and wildlife habitats frequently deprived invaded human communities of their primary means of subsistence, while their 'resources' were exploited for imperial profit. The importation of sheep and cattle, cash-cropping and other European agri-practices replaced local hunting and gathering or systems of crop rotation, thereby damaging established ecosystems, reducing soil fertility, and even, as in the case of African settlements south of the Sahara, causing famine through desertification.

Whatever the initial impact or the long-term legacies of incursion, the dispossessed frequently faced poverty, illness and starvation and the origi-

---

[3] Buell, *The Future of Environmental Criticisms*, 97.

[4] Richard H. Grove, *Ecology, Climate and Empire: The Indian Legacy in Global Environmental History, 1400–1940* (New Delhi; Oxford UP, 1998): 3.

nal accommodated relations between environment, other animals and humans were fractured beyond restitution. Such widespread disruptions of indigenous practices and conceptions have been the most significant and long-lasting consequences of the post-1492 conquests and invasions. Communal biotic subjectivities and place-integrated identities were irretrievably suborned by a European instrumentality yoked to a destructive ideology of technological 'progress' that was rapidly applied to the 'new' land and its inhabitants. Writing of landscape painting, W.J.T. Mitchell observes that semiotic features of landscape, and the historical narratives they generate, are tailor-made for the discourse of imperialism, which conceives itself precisely (and simultaneously) as an expansion of landscape understood as an inevitable, progressive development in history, an expansion of 'culture' and 'civilisation' into a 'natural' space in a progress that is itself narrated as 'natural'. Empires move outward in space as a way of moving forward in time; the 'prospect' that opens up is not just a spatial scene but a projected future of 'development' and exploitation.[5]

For many non-Western peoples, the very concept of 'the human' had been constructed by and within their surrounding environments. Where Westerners apprehended relations between themselves and 'their' land as one of ownership (or, at best, stewardship), many other peoples understood their humanness as constituted and expressed *through* it, rather than, as in post-Enlightenment Western philosophies, *against* it: place was not so much crucial to identity as actively constituent of it. The modern West, it is said, exists *in relation* to nature; other peoples had often existed *as* 'nature'. Such conceptions of human identity did not usually function within imperial / colonial encounters as simple alternatives or tolerated differences; their relationship to each other was necessarily closer and more destructive. Enlightenment reification of Reason reinforced earlier Western notions of nature and, in particular, cast 'wilderness' as the antitheses of 'civilization'. The closer to nature a people, the less deserving of ethical concern they were. Nature conceived of as the antithesis of the human (or human culture) was thus a 'naturally' inferior realm inhabited by 'savage' species of all kinds. These "areas of rational deficit" (as the philosopher

[5] W.J.T. Mitchell, "Imperial Landscape," in *Landscape and Power*, ed. W.J.T. Mitchell (Chicago: U of Chicago P, 1997): 17.

Val Plumwood puts it)[6] could be considered 'empty' by the West, free for European taking; while the settlers' or administrators' instrumental views and practices generally precluded persistence of local alternatives.

Paradoxically, however, as the antithesis of the 'civilized', the 'savage' was itself philosophically essential to Western self-constitution. The Western definition of humanity depended (and still depends) on the presence of the not-human – the uncivilized, the animal and animalistic.[7] Western justification for invasion and colonization proceeded from this basis, and, as Val Plumwood has argued, understood non-European lands and the people and animals who inhabited them as "spaces," "unused, underused or empty" (53). The very ideology of colonization is thus one where anthropocentrism and eurocentrism are inseparable: the "anthropocentrism that underlies Eurocentrism justified those forms of European colonialism which see indigenous cultures as 'primitive', less rational, and closer to children, animals and nature."[8] Within many cultures – and not just Western ones – anthropocentrism had long been 'naturalized'. Absolute prioritization of one's own species interests over those of the silenced majority is regarded as 'only natural'. Ironically, of course, it is precisely through such appeals to 'nature' that other animals and the environment are excluded from the privileged ranks of 'the human', rendering them available for our exploitation. As Cary Wolfe (citing Jacques Derrida and Georges Bataille) puts it,

> the humanist concept of subjectivity is inseparable from the discourse and institution of [a] speciesism which relies on the tacit acceptance that the full transcendence of the human requires the sacrifice of the animal and the animalistic, which in turn makes possible a symbolic economy in which we can engage in a 'non-criminal putting to death', as Derrida phrases it, not only of animals but of humans as well by marking them as animal.[9]

---

[6] Val Plumwood, "Decolonising Relationships with Nature," in *Decolonizing Nature: Strategies for Conservation in a Post-Colonial Era*, ed. William M. Adams & Martin Mulligan (London: Earthscan, 2003): 52.

[7] See Jacques Derrida, "' Eating Well' or the Calculation of the Subject: An Interview with Jacques Derrida," in *Who Comes After the Subject?* ed. Eduardo Cadava, Peter Connor & Jean–Luc Nancy (New York & London: Routledge, 1991): 96–119, and Derrida, "Force of Law," *Cardozo Law Review* 11 (1990): 919–1045.

[8] Plumwood, "Decolonising Relationships with Nature," 53.

[9] Cary Wolfe, "Old Orders for New: Ecology, Animal Rights and the Poverty of Humanism," *Diacritics* 28.2 1998: 39.

The effectiveness of this "discourse of species" is that "when applied to social others of whatever sort", it relies upon "the taking for granted of the institution of speciesism; that is, upon the ethical acceptability of the systematic, institutionalized killing of non-human others."[10] In other words, in *assuming* a 'natural' prioritization of humans and human interests over those of other species on the earth, we are repeating the racist ideologies of imperialism on a planetary scale. In working towards a genuinely post-imperial, environmentally based conception of community, then, a re-imagining and reconfiguration of the human place in nature necessitates an interrogation of the category of the human itself and of the ways in which the construction of ourselves *against* nature (with its attendant hierarchization of other life forms) has been and is complicit in colonialist and racist exploitation from conquest to the present day.

Postcolonial studies, ethically concerned and increasingly conscious of historical (and persisting) environmental considerations, has come to understand environmental issues as not only focal in the European conquest and colonialization of the globe, but as inherent in the very ideologies of imperialism and racism on which such invasion and colonialism depended. Not only were other peoples regarded as part of 'nature' (and thus treated purely instrumentally as animals), they were also forced into or co-opted over time to Western views of the environment, thus rendering both cultural and environmental restitution difficult if not impossible. Once invasion and 'settlement' had been accomplished (or administrative structures put in place), the environmental impacts of Western attitudes to 'human being' in the world were further facilitated or reinforced by the deliberate (and accidental) transport of animals, plants and peoples throughout the European empires, instigating widespread ecosystem change under unequal power regimes. (Such primary transformations frequently laid the foundations of today's international trade patterns whose inequalities in design and implementation are often euphemized as a new (and positive) 'globalization'.)

Writing of the relatively recent field of environmental history, Donald Worster nominates what he regards as the major "targets" for rendering environmental considerations "central"[11] to the discipline of History as a whole: since environmental history is "where the natural and cultural inter-

---

[10] Wolfe, "Old Orders for New," 39.

[11] Donald Worster, "Transformations of the Earth: Toward an Agroecological Perspective in History," *Journal of American History* 76.4 (March 1990): 1087–1106.

sect and interact with each other," it becomes "increasingly difficult for most historians to force [environmental issues] to the periphery of historical concerns."[12] A greater knowledge of the structure and distribution of natural environments in the past "is essential, followed by studies of the ways in which technology has restructured human ecological relations (the way people have tried to make nature move over into a system that produces resources for their consumption)."[13] Last (but not least) of Worster's "targets" for the discipline as a whole is examination of "that more intangible, purely mental type of encounter in which perceptions, ideologies, ethics, laws and myths have become part of an individual or group's dialogue with nature"; such "patterns of human perception, ideology and value," Worster notes, "have often been highly consequential, moving with all the power of great sheets of glacial ice."[14] These three areas, then, should form the basis of historical inquiry and scholarship *as a whole*, and investigation of the connections between them will, Worster predicts, catalyze the new and needed perspectives.

As a less cohesive discipline (both conceptually and methodologically), "English" in general and postcolonial studies in particular have yet to resituate environmental concerns at the very centre of their disciplinary inquiries. Nevertheless, for postcolonial studies, examination of this interface between 'nature' and 'culture' is pertinent and increasingly urgent. Postcolonialism's concerns with conquest and colonization; with race; with the imposition (and, more rarely, 'exchange') of cultural knowledge; its investment in theories of indigeneity and diaspora and of conceptions of and relations between native and invader are also the central concerns of animal and environmental studies. Moreover, as Worster acknowledges, it is in the myriad relationships between material practices and ideas (especially in cross-cultural contexts) that day-to-day planetary life is lived and futures governed. Within human cultures, however, ideas *and* practices are inseparable from issues of representation.

In *The Columbian Exchange* and *Environmental Imperialism*, Alfred Crosby considered the ways in which both ideas and materials were "exchanged" between old world and new on anything but level playing fields. In the colo-

---

[12] Richard White, "Environmental History, Ecology and Meaning," *Journal of American History* 76.4 (March 1990): 1111.

[13] Worster, "Transformations of the Earth," 1091.

[14] "Transformations of the Earth," 1091.

nies of occupation, these radical inequalities or "exchanges" seemed most evident – at least initially – in the military and political spheres, while in the settler colonies it was the results of *environmental* imperialism which were often most immediately evident. Different conceptions of being-in-the-world had indeed been 'exchanged' by individuals or groups in colonialist circumstances: Eastern religions had long intrigued Europeans, and the oral cultures of the South Pacific and Africa provoked interest and admiration in many explorers. But in Australia, North America, New Zealand and South Africa, genuine curiosity about and respect for indigenous cultures, philosophies and religions was rare. Even the most well-intentioned missionaries, settlers and administrators conceived of themselves as conferring / imposing the 'gifts of civilization' upon the benighted heathen with little or no interest in receiving his or her philosophical gifts in return. Settlers arrived with crops, flocks and herds, and 'cleared' land, exterminating local ecosystems, while human, animal and plant 'specimens' taken to Europe from these 'new' worlds were, by contrast, few and often in inert form. (No human, animal or plant, whether wild or domesticated, transported from the colonies to Europe was in any position to wreak comparable havoc on European ecosystems.) Moreover, they did not arrive as part of traditional agricultural or pastoral practices or with the authority of the normative: They were, instead, isolated 'exotics', as Renata Wasserman observes:

> Indians paraded before royal courts, like turkeys and parrots in cages were the innocent signifiers of an otherness that was [...] exotic, that is, non-systematic, carrying no meaning other than that imposed by the culture to which they were exhibited.[15]

By contrast, European imports to the newly settled colonies – humans, animals, plants – were regarded by the (soon dominant) invader-cultures as necessary and 'natural' impositions on or substitutes for the local wilderness or "bush"; and even if these invading species were initially difficult to establish or acclimatize, they soon prospered in lands where their control predators were absent. The genuinely 'natural' ways of indigenous ecosystems were irretrievably undone as 'wild' lands were cleared for farming or 'opened up' to pastoralism.

This book considers these unbalanced environmental "exchanges" within the context of British imperial power and colonial rule, as well as

---

[15] Renata Wasserman, "Re-Inventing the New World: Cooper and Alencar," *Comparative Literature* 36.2 (1984): 132.

charting some contemporary environmental legacies of those (still) inequitable imports and exports, transportations and transmutations. Skewed (especially in the settler colonies) in favour of the colonizing (human) culture, such "exchanges" were nevertheless often more complex in practice than this apparently 'simple' pattern of invasion, land-clearing, and destruction (or erosion) of indigenous ways and ideas might lead us to suppose. And, as Worster suggests, the material and ideological (which go hand in hand) are often disturbed and / or problematized by the complexities and persistence of the local they displace or by the inappropriate and contradictory ideas and practices they catalyze. European / Western ideas of human environmental relations transported to the colonies undergo pressures which are sometimes transformative in their turn. Inefficacious as they might often have seemed, the indigenous peoples, as well as animals and plants of the colonies, sometimes altered, albeit to a limited degree, European Self / Other conceptions and practices. More usually, however, ideas of land-use formed in Britain and Europe predisposed colonial administrators and settlers to an easy belief in the apparently limitless farmland of the settler colonies. Such places were, after all, apparently untamed, un-owned and, above all, *unused.* Settlers thus set about rendering them productive and profitable through imported methods rather than by accommodating to local circumstances.

Leigh Dale's opening essay in this volume considers the ways in which sheep farming in Australia contributed greatly to the transformation of large tracts of land as it dispossessed the native inhabitants. As unwilling / unwitting 'agents' of the colonizing Europeans, sheep were central to 'land clearing' (in every sense of the term) and also became representational scapegoats for their human importers, as if they were the primary agents of environmental change. At the same time as they were thus made proxy for the ecological damage of empire, the sheep also served as a symbol of (new, white) nationhood. That such massive environmental transformation could be disconnected from settler agency through a process of representation was not peculiar to Australia. As J.M. Coetzee's white South African narrator, Magda, puts it in *The Heart of the Country:*

> Fascinating, this colonial history [...] And economics: how am I to explain the economics of my existence, [...] unless the sheep have something to eat (for this is not finally an insect farm) [...] It must be the scrub that nourishes the sheep that nourish me [...] There is another great moment in colonial

> history: the first merino is lifted from shipboard [...] unaware that this is the promised land.[16]

Through the use of the passive voice in this passage, ("the first merino is lifted ..."), settler agency and responsibility are elided. The sheep simply arrived, changing landforms and destroying or (literally) unsettling indigenous cultures.

Chris Tiffin and Ruth Blair also consider the ways in which the transportation of plants (and animals) into and across the Pacific in the nineteenth century – even where intentionally benign or merely nostalgic – had widespread destructive effects on invaded environments. Many Europeans had envied the 'simpler' life in "the South Seas" since the eighteenth century, but, as traffic across the ocean increased, the endemic or indigenous was, as Blair notes, rarely appreciated for its own sake. Plants were valued only as either "potential commodities or for aesthetic qualities that fitted into the picture of the South Sea paradise that early explorers had constructed and promulgated in Europe." For, as W.J.T. Mitchell has noted,

> imperialism is clearly not a simple, single, or homogeneous phenomenon but the name of a complex system of cultural, political, and economic expansion and domination that varies with the specificity of places, peoples, and historical moments. It is not a "one-way" phenomenon but a complicated process of exchange, mutual transformation and ambivalence. It is a process conducted simultaneously at concrete levels of violence, expropriation, collaboration, and coercion, and at a variety of symbolic and representational levels whose relation to the concrete is rarely mimetic or transparent.[17]

While importation of sheep and cattle on the scale Dale describes had a larger impact on indigenous ecosystems, some of the transportation of plants and animals which proceeded from more convoluted motives could also have widespread and unexpected results. Colonists, missionaries, hunters, naturalists and explorers (who transported specimens – dead and alive – from the colonies back to Britain and Europe) were motivated by 'normative' European ideas: by nostalgia for European plants, animals and landscapes; or by a quasi-theological impulse to re-create the "Lost Garden," in all its former fullness, by regrouping the species apparently scattered after the Fall. A similar impulse lay behind the opposite practice: "To distribute the good things of the earth throughout the world," As

---

[16] J.M. Coetzee, *In the Heart of the Country* (1977; Harmondsworth: Penguin, 1982): 19.

[17] W.J.T. Mitchell, ed. *Landscape & Power*, 9.

Chris Tiffin notes in the essay from which this volume takes its name, the Queensland (Australia) Acclimatisation Society in 1865 listed its disbursements of Australian fauna and flora to other parts of the globe and its importation of exotic species as follows: "'Two kangaroos to the Royal Society in London, two scrub turkeys to New Caledonia, fish from the Mary River to Tasmania, and 'five emus to the King of Siam'." While the emus were unlikely to affect South East Asian ecosystems, not all such exchanges were as harmless in intent or effect. The report also listed the importation into Australia of "red deer" distributed to a "local landholder" for breeding purposes. Although in terms of environmental impact such "exchanges" pale into insignificance beside large-scale pastoralism, the ideologies they both represented and helped to inculcate could have, like some of their own imports, disastrous consequences. While the impulse behind Acclimatisation Society imports and exports was primarily the distribution of "the good things of the earth," members justified their applications for government grants by pointing to the ways in which their activities had often provided the basis of colonial economies. When the nascent Queensland Acclimatisation Society petitioned the Colonial Legislative Council for a grant, it was able to argue validly that

> 'The chief producing interest in the colony owes its origin to the importation of animals not indigenous to Australia, and that almost all future products, such as sugar, coffee, tobacco, cotton, etc., whether for home consumption or exportation will have similar orginations'."

Such food crop imports, as Tiffin shows, were sometimes "startlingly experimental."

Under the influence (or edicts) of missionaries as well, accommodated ecocultural relations were often redefined and recast. European ideas of family living and environmental order, sometimes including the growing of Euro-domestic flora, altered native landscapes and introduced plants which became noxious weeds in new locations. (Jack Goody, for instance, has noted that Christian converts' houses in North Africa could be identified immediately by the floral cultivation surrounding their homes.) More damagingly, the veneration of totemic animals whose hunting had usually been scrupulously regulated to ensure their survival and that of their human predators was debunked as Christian ideas of animal 'use' – that is, their "unbridled" commodification – triumphed. Spatial arrangements and notions of land ownership were also reconfigured. As Anna Johnston shows, "planting the seeds of Christianity" was much more than a metaphor for human conversion. In aspiring to turn Polynesia into "a Pacific version of a Birmingham suburb," missionaries destroyed or destabilized local land-use patterns, interfered with ecosystems, disrupted the bases of their converts' former ontologies, and sometimes also annihilated their traditional means of subsistence. The Rev. Barff reported on his 1830 visit to Maiasiti that "a larger number of neatly plastered houses had been erected. Each house was surrounded by a garden, well stocked with potatoes and other vegetables, and [...] the whole settlement presented the appearance of a neat village, in a state of great prosperity. Order, harmony and industry prevailed."

James Anthony Froude, on his late-nineteenth-century post-emancipation tour of British West Indian territories, laments the "lapsing" into "jungle" of Jamaica's once ordered and cultivated fields of the slave era. Worse, in Froude's view, was the lack of enterprise demonstrated by the current British generation, who, without the imperial energy of their forebears, were prepared to leave so much of the island of Dominica "unimproved," neglecting "a colony which might yield them wealth beyond the treasures of the old sugar planters." Like Trollope, Froude regarded the condition of Barbados as being in exemplary contrast to Jamaica. Once "covered with forest," Barbados was, by the late-nineteenth century, devoid of all except "cabbage palms and detached clumps of mangy-looking mahogany trees"; and, the "jungle" having been disposed of, "human beings have taken its place." The "garden" island of Barbados was neither gone to seed *nor* "underdeveloped," since slavery and continued careful cultivation had ensured that all was well to the European traveller's eye.

Relationships between the concepts of wilderness and garden are, as Carolyn Merchant argues in the American case, very complex ones, inevitably underpinned by the Biblical mythology of the Fall and the expulsion of humans from Eden to wander in "the wilderness".[18] Labour in the earth – cultivation – was the only road to recovery. But in the Caribbean the potential righteousness of labour was already contaminated by its perversion in slavery, hence the relationship between garden cultivation and original wilderness (or "jungle" – the word imported by the British from

[18] Carolyn Merchant, "Reinventing Eden: Western Culture as a Recovery Narrative," in *Uncommon Ground: Rethinking the Human Place in Nature*, ed. William Cronin (New York: W.W. Norton, 1996): 132–70.

India) was a particularly complex and troubled one. Nevertheless, throughout Britain's Empire, imperial ideals and colonial practices sponsored the destruction of primal forest and its replacement with human tilling to convert 'empty' space to profitable 'place'. Settler-invaders were not, from their perspectives, murdering the inhabitants and destroying indigenous ecosystems; they were rendering their 'new' land 'fruitful'.

Ideas and material practices are intrinsically interwoven, as in the 'modernity' of colonialist cultivation and the Victorian ideology of 'progress'. But while both eighteenth- and nineteenth-century European conceptions underpinned environmental damage and destruction in colonized lands, they did not always themselves remain unaffected. Radical environmental differences and local exigencies forced practical changes *and* energized new ways of thinking, even if these were sometimes rationalizations of settler colonialist ideologies and practices. Transported methods, philosophies and genres, in exchanging locales, were sometimes subtly altered or had to be adapted to different purposes in unfamiliar territory.

Landscape painting, as Catherine Howell argues, "forms a crucial representational and symbolic site on which Australian colonialism cultivated its terrain." In her discussion of the "relationships between colonialism, aesthetics and cultural identities," Howell uses the example of the work of the nineteenth-century British-Australian painter John Glover to demonstrate the ways in which the complex and contradictory attitudes of the colonial settler to the Australian environment were reflected in representations of the landscape. Her essay also offers a timely reminder that colonialist "exchanges" cannot always be read in the direction of the power dynamic; and she stresses the need for current theoretical and literary critical practice to be more self-aware and critical (especially within postcolonial studies), where there is too often "a tendency to construct the past as deluded and the present as enlightened."

Andrew McCann also examines the transmutation of a key European aesthetic movement – Romanticism – from its eighteenth- and nineteenth-century genesis and developments in England and Europe to the colonies. Particularly in the settler colonies – McCann's example is Australia – "the refusal of instrumentality in Romantic elegy is compromised by the complicit forces of colonisation and commodification," so that, in its transfer to such colonies, the "notion of nature as object of lament" so important in European Romanticism "gives way to representation of landscape in which resistance to instrumentality becomes increasingly difficult to express." McCann also addresses two major theoretical concerns of studies in the environmental humanities as a whole: first, the ways in which the "normative impulse that circulates around our need for nature" was necessarily deflected, by settler-colony sociocultural circumstances, into discourses and "representational practices beholden to instrumentality." In tackling this complex shift, he also considers the always vexed relationship between representation and 'the real', concluding that, if 'nature' cannot be apprehended by humans aside from our representations of it, we can nevertheless still talk about "*the ethical implications* of different representational practices." By so doing, McCann argues, we may be able to "at least imagine the possibility of a semiotics of nature adequate to a progressive, ecologically conscious politics."

This volume is concerned with empire and environment in both the nineteenth and the twentieth century, and two chapters concerning the first half of the twentieth century are indicative of important value-shifts. While imperial / colonial transfer and "exchanges" still take place on the uneven field of European imperialism and colonization (with their commodification of 'nature'), in the late-nineteenth century a revaluation of the indigenous (places, people, fauna and flora), as against the invader-imported, began to be energized by an increasing apprehension of the possibility of the actual extinction of unique humans, animals and ecosystems. Richard Grove argues, *contra* the arguments for settler environmental destruction, that it was in Europe's colonies that the concept of conservation in the form of National Parks was first mooted and implemented.[19] In Europe, environmental degradation under human population pressures had been a longer and slower process, one which had merely been accelerated by the Industrial Revolution. But in the colonies, particularly, again, the settler colonies, the transformation of "jungle" or "wilderness" into farm or pastoral land was rapid, widespread, and observably destructive of the indigenous environment. The annihilation (or near-annihilation) of human races (for example, the Caribs in the West Indies) and the decimation or extinction of once numerous animals and birds (such as the buffalo and the passenger pigeon in the Americas) had brought 'home' the knowledge that species-disappearance might not be the province of geological time alone. While indigenous peoples and ecosystems were usually

[19] Richard H. Grove, *Green Imperialism: Colonial Expansion, Tropical Island Edens, and the Origins of Environmentalism, 1600–1860* (Cambridge: Cambridge UP, 1995).

politically and biologically 'powerless' in the face of European occupation or settler-colony invasions, the late nineteenth and early twentieth centuries began to usher in – on the part of at least some Europeans – a greater interest in and respect for other cultures and places as having intrinsic value beyond their potential for commercial exploitation.

Thus, as Carrie Dawson shows, an Englishman in Canada, Archie Belaney (later famously known as "Grey Owl"), did not so much 'go native' in the older, European colonialist sense (where to do so was to cast oneself beyond the 'civilized' pale in the eyes of fellow Europeans or settlers of European descent) as live out a strategy of conservation of animals and ecosystems *through* the adoption of nativist values attuned to specific place and the preservation of its extra-human inhabitants. In a very different register, Jo Robertson's essay also covers a crucial period of revaluation and redeployment. In a number of British colonies, leprosaria had been located on offshore islands to contain the disease. When *Bacillis leprae* was finally isolated in the late-nineteenth century and both the agent of transfer *and* its low communicability established, leprosaria were converted to political prisons for independence agitators, the most famous of which, Robben Island, had long incarcerated the man who would become the first president of post-apartheid South Africa, Nelson Mandela. Jo Robertson has explored elsewhere the ways in which diseases (and the "cameraderie" induced by paranoid fears of diseased 'Others') became (like sheep) improbable agents in nation-formation.[20] But in her essay in the present volume the focus is on the various transformations of a particular environment, Peel Island off the Southern Queensland coast, from convict prison to leper colony, to (proposed) tourist resort. As political prisoners went on to become Independence leaders, the colonial locations that leprosaria and political prisons had (somewhat ironically) preserved sometimes became conservation parklands.

Tourism, as the Antiguan writer Jamaica Kincaid demonstrates, is a 'natural' successor to imperial invasion. Westerners (and the urban rich generally) travel to observe exotic environments, 'bringing home' the (new) products of their journeys – souvenirs, photographs, anecdotes and, all too frequently, re-animated stereotypes of poorer peoples and (apparently) 'untouched' ecoscapes. In *A Small Place*, Kincaid interrogates the terms of

---

[20] Jo Robertson, "In a State of Corruption: Loathsome Disease and the Body Politic" (doctoral dissertation, University of Queensland, Australia, 1999).

contemporary tourism,[21] and demonstrates the persisting inequities of this imperialist / tourist "exchange"; one whose continuing exploitations of post-independence colonies like Antigua are necessarily imbricated with domestic political corruption through the legacies of imperial history and the current monetary demands of international bodies like the IMF. Tourism thus becomes a "necessary evil" for many formerly colonized countries, producing, as Arundhati Roy and Amitav Ghosh have noted, continuing (if different) clashes between the local and the international. People forced off their lands to make way for game parks and reserves patronized by wealthy white tourists are naturally unsympathetic to (Western) government-sponsored 'conservation' measures which exile them from their traditional ecosystem and subsistence bases.[22]

Just as sheep became, through the ways in which they were represented, colonialist 'scapegoats' for settler-colony environmental destruction, and subtle shifts in colonial Romanticism facilitated the commodification of nature (rather than a sharpened perception of its passing), so late-twentieth-century travellers in Africa, as Gillian Whitlock argues, use Karen Blixen's *Out of Africa* to perform a self-justifying erasure of the colonialist history of Kenya. In "imagining a past – Kenya's – as representative for the entire continent, the quite uncharacteristic and specific nature of Kenya's recent and relatively short-lived existence as an elite settler colony in British East Africa is obscured," eliding in the process the European destruction of a number of Africa's ecosystems. Significantly, the 'trick' is again effected through reading and representation.

While acknowledging that ecotourism certainly appears to offer more ethical options to the contemporary tourist than many other forms of travel, Helen Gilbert demonstrates the ways in which both its advertising and tours often replicate, albeit in apparently more benign form, the old relations of Empire and colony as well as sharing many of the drawbacks of the more established kinds of tourism. In his discussion of the Sandline Mercenaries Affair, Robert Dixon considers a much more sinister contemporary invasion of Papua New Guinean democracy and Bougainville sovereignty, tracing through an almost breathtaking series of events the ways in which the possession of natural resources and the pressures of interna-

---

21 Jamaica Kincaid, *A Small Place* (1988; Harmondsworth: Penguin, 1989): passim.

22 Jody Emel & Jennifer Wolch, "Witnessing the Animal Moment," in *Animal Geographies: Place, Politics and Identity in the Nature – Culture Borderlands* (London: Verso, 1998): 20.

tional finance (together with the legacies of colonialism) force beleagured post-independence governments into further unequal and destructive liaisons with outside bodies to produce even more communal problems and environmental degradation. Former colonies may have regained their independence from Britain, but they remain prey to international monetary games played in places well beyond local control.

A different clash between the Indo-traditional and the Euro-modern (in this specific case, 'scientific') assumptions and ways of knowing and understanding is explored in Meenakshi Sharma's essay on the debates over the purity (or otherwise) of the waters of the Ganges. Here two apparently incommensurable approaches – the spiritual and the scientific/materialist – necessarily clash. In contrast to the situation in a number of other formerly colonized cultures, in India traditional attitudes have not necessarily been forced out of court by the Euro-scientific, and Sharma's essay shows how Indian *and* Anglo-European attitudes continue to exist side by side, with both approaches encouraging and inhibiting (in their very different ways) the attempted implementation of conservationist measures.

Susie O'Brien's chapter steps back from the specific environmental issues themselves to consider the ways in which some of the assumptions and practices of early 'ecocritics' and 'ecocriticism' – most commonly associated with the US academy – inadvertently replicate colonialist attitudes, albeit in an environmentally positive and well-intentioned practice. We have generally avoided using the term 'ecocriticism' here, both because of its American neocolonialist potential to dominate this burgeoning contemporary field and because the present contributions – while they stress the inescapable importance of representation – are determinedly interdisciplinary.

In this volume a variety of significant connections between Empire (in this case the British Empire) and 'environment' in its broadest, ecosystemic sense are explored in both the nineteenth and twentieth centuries and in the settler colonies as well as the colonies of occupation. If the key theme is transfer and/or exchange within contexts of unequal power relations (both *between* peoples and between humans, animals and extra-human environments), the "exchange" required to resituate environment within a post-imperial ecologically conscious matrix is a fundamental reconfiguration of the very idea of 'community' to embrace not just the formerly imperial and the formerly colonized (on new and equitable bases), but also human and extra-human 'voices'. The radical political philosopher Bruno Latour notes that "ecology movements have sought to position themselves on the political chessboard without redrawing the squares, without redefining the rules of the game, without redesigning the pawns"; for Latour and others, such primary "conceptual work" is "indispensable" in any attempt to address local or global ecological crises.[23] Such foundational inquiries, Latour continues, should move us closer to a [re] "convoking [of] the collective [...] on radically different grounds." But this, as he points out, is a far more complex task than simply arriving at a "reunification of things and people, objects and subjects." Among other philosophical shifts, it involves "adding a series of new voices to the discussion, voices that have been inaudible up to now."[24] To allow other, non-human voices to speak is also to initiate a thorough investigation of the complex relationships we have created or denied between what we categorize as 'facts', and what we understand as 'values':

> The aberrant opposition between mute nature and speaking facts was aimed at making the speech of scientists *indisputable*, thus, this speech passed through a mysterious operation resembling ventriloquism from "I speak" to "the facts speak for themselves" to "all of you have to do is shut up."[25]

This in turn requires a reconsideration of the ways which we represent ourselves 'as' or 'in' a 'nature' which the dominant West has historically alienated from human being through wholesale commodification in the service of capitalist 'progress'. We like to believe that a corollary of such rampant 'progress' has been an ethical advance towards more inclusive notions of 'community' and more equitable patterns of cultural and material exchange. But until the concept of community, for the West particularly, embraces the extra-human, forms of racism and imperialism will inevitably persist, since all are inherently tied to our persisting – 'natural' – investment in species dominance and the unquestionable priority of (Western) interests.

WORKS CITED

Buell, Lawrence. *The Future of Environmental Criticism* (Oxford & Malden MA: Blackwell, 2005).

Coetzee, J.M. *In the Heart of the Country* (1977; Harmondsworth: Penguin, 1982).

---

23 Bruno Latour, *Politics of Nature: How to Bring the Sciences into Democracy*, tr. Catherine Porter (Cambridge MA: Harvard UP, 2004): 5, 3.

24 *Politics of Nature*, 69.

25 *Politics of Nature*, 68.

Crosby, Alfred W. *The Columbian Exchange: The Biological and Cultural Consequences of 1492* (Westport CT: Greenwood, 1972).

——. *Ecological Imperialism: The Biological Expansion of Europe, 900–1900* (Cambridge: Cambridge UP, 1986).

Derrida, Jacques. "'Eating Well' or the Calculation of the Subject: An Interview with Jacques Derrida," in *Who Comes after the Subject?*, ed. Eduardo Cadava, Peter Connor & Jean–Luc Nancy (New York & London: Routledge, 1991): 96–119.

Derrida, Jacques. "Force of Law," *Cardozo Law Review* 11 (1990): 919–1045.

Emel, Jody, & Jennifer Wolch "Witnessing the Animal Moment," in *Animal Geographies: Place, Politics and Identity in the Nature – Culture Borderlands* (London: Verso, 1998): 1–24.

Ghosh, Amitav. *The Hungry Tide* (London: HarperCollins, 2004).

Goody, Jack. *The Culture of Flowers* (Cambridge: Cambridge UP, 1993).

Grove, Richard H. *Ecology, Climate and Empire; The Indian Legacy in Global Environmental History, 1400–1940* (New Delhi: Oxford UP, 1998).

——. *Green Imperialism: Colonial Expansion, Tropical Island Edens, and the Origins of Environmentalism, 1600–1860* (Cambridge: Cambridge UP, 1995).

Kincaid, Jamaica. *A Small Place* (1988; Harmondsworth: Penguin, 1989).

Latour, Bruno. *Politics of Nature: How to Bring Science into Democracy*, tr. Catherine Porter (Cambridge MA: Harvard UP, 2004).

Merchant, Carolyn. "Reinventing Eden: Western Culture as a Recovery Narrative," in *Uncommon Ground: Rethinking the Human Place in Nature*, ed. William Cronin (New York: W.W. Norton, 1996): 132–70.

Mitchell, W.J.T., ed. *Landscape and Power* (Chicago: U of Chicago P., 1997).

Plumwood, Val. "Decolonising Relationships with Nature," *PAN* 2 (2002). Repr. in *Decolonising Nature: Strategies for Conservation in a Post-Colonial Era*, ed. William M. Adams & Martin Mulligan (London: Earthscan, 2003): 51–78.

Robertson, Josephine. "In a State of Corruption: Loathsome Disease and the Body Politic" (doctoral dissertation, University of Queensland, 1999).

Roy, Arundhati. *The Cost of Living* (London: HarperCollins/Flamingo, 1999).

Wasserman, Renata. "Re-Inventing the New World: Cooper and Alencar," *Comparative Literature* 36.2 (1984): 129–43.

White, Richard. "Environmental History, Ecology, and Meaning," *Journal of American History* 76.4 (March 1990): 1111–16.

Wolfe, Cary. "Old Orders for New: Ecology, Animal Rights and the Poverty of Humanism," *Diacritics* 28.2 (1998): 21–40.

Worster, Donald. "Transformations of the Earth: Toward an Agroecological Perspective in History," *Journal of American History* 76.4 (March 1990): 1087–1106.

Empire's Proxy
Sheep and the Colonial Environment

LEIGH DALE

THE RESOURCES AVAILABLE TO PASTORAL EXPANSION in the nineteenth-century Australian colonies were said to be limitless: as Kevin Frawley, among many others, points out, "the most persistent image [of the land] is of inexhaustibility."[1] But experience was to prove that in a relatively short space of time the rich pastoral resources could be dramatically diminished or destroyed by the colonizers. Geoffrey Bolton declares that, given the fact that by 1860 the country was carrying twenty million head of sheep and four million of cattle, and by 1890 a hundred million sheep and eight million cattle,

> None of the exotic invaders that ran wild across the country, not even the rabbit, brought about such profound transformation of the Australian environment as beef cattle and merino sheep. [...] Their spread was accomplished in a hundred years, and in that space of time the original [...] bush gave way to a landscape and environment created very largely in the interests of the flocks of sheep, the herds of cattle and the men and women whose economy depended on them.[2]

---

[1] Kevin Frawley, "Evolving Visions: Environmental Management and Nature Conservation in Australia," in *Australian Environmental History: Essays and Cases*, ed. Stephen Dovers (Oxford: Oxford UP, 1994): 63.

[2] Geoffrey Bolton, *Spoils and Spoilers: A History of Australians Shaping Their Environment* (Australian Experience series; Sydney: Allen and Unwin, 2nd ed. 1992): 81.

This invasion of European animals and their owners had a devastating effect on the indigenous peoples, on the land and its vegetation. In this essay I want to examine just a few of the many texts in which sheep are represented, and to consider their figuration as agents of major environmental change and as active instruments in the appropriation of land. The text I will be focusing on is C.E.W. Bean's popular history *On the Wool Track*, first published as a collection in 1910.

Sylvia Hallam makes the point that Aboriginal land-care practices were crucial to the creation of the vast expanses of grasslands that so intoxicated squatters.[3] This kind of argument, when it was first made by academics some thirty years ago, represented a major rethinking of white Australians' understanding of Aborigines' relationship to the land, now dignified by the vocabulary of capitalism: Aborigines have become managers. A second, again seemingly recent, change to ways of thinking about landscape has been the promotion of views about the fragility of the Australian soils and vegetation, a view expressed with particular eloquence in the writing of Judith Wright and Eric Rolls.[4] This reconstruction of the Europeans as environmentally destructive has now become the orthodox view; thus, Tom Griffiths comments that when the British arrived on "the Australian pastoral frontier" it provided them "and their flocks with a short-lived bounty, an ecological niche that was exhausted in their lifetimes."[5] It should be noted, though, that this is a view contested by contemporary writers like Ann Young who suggest that claims about environmental damage have been exaggerated, and Timothy Flannery, whose claims that Aborigines caused the extinction of the megafauna have themselves drawn strong criticism.[6] Given this controversy over the ecological impact of colonization, it is useful to turn back to historical and literary accounts of pastoral expansion to examine nineteenth-century ways of thinking about sheep, colonialism, and environment.

---

[3] Sylvia Hallam, "The First Western Australians," in *A New History of Western Australia*, ed. C.T. Stannage (Perth: U of Western Australia P, 1981): 64.

[4] Judith Wright had, of course, been making this argument for many decades. See Sandra Brunet, "Landscape and Identity: Judith Wright's Fragile Land and Haunted Self" (unpublished MA thesis, University of Queensland, 1998).

[5] Tom Griffiths, intro. to *Ecology and Empire: Environmental History of Settler Societies*, ed. Tom Griffiths & Libby Robin (Edinburgh: Keele UP, 1997): 11.

[6] Ann R.M. Young, *Environmental Change in Australia since 1788* (Melbourne: Oxford UP, 1996); Timothy Flannery, *The Future-Eaters: An Ecological History of the Australasian Lands and Peoples* (Port Melbourne: Reed, 1994).

The 1820s saw the beginning of rapid colonization of inland Australia, with the emigration of larger numbers of free settlers and the completion of the road over the Blue Mountains, and the town of Bathurst established in 1815. This was a period in which a transformation in thinking about the Australian colonies began to take place: specifically, about the capacity of the land and rural industries to provide profits for British capital – a shift both signalled and encouraged by the Bigge Report of 1821 on the state of the colony of New South Wales. The status of sheep and the wool industry, frequently organized as a narrative of prosperity and progress that has as its foundation the life and work of John Macarthur, was central to this shift in thinking. A sign of the change, and of resistance to it, can be found in Peter Cunningham's 1827 publication *Two Years in New South Wales.* While Cunningham claimed that any mention of the colony to a stranger almost inevitably prompted a comment on wool, a reviewer in *Blackwood's Magazine* disputed the assertion, contending that "for one man in whom the name [New South Wales] calls up the idea of Merino wool, there are ninety-nine [...] who think only of ropes [...], arson, burglary, kangaroos, George Barrington, and Governor Macquarie."[7] Cunningham's claim about the importance of wool signals its foundational status in the economic *and* cultural shift from 'penal colony' to 'independent economy,' specifically the centrality of 'the wool industry' to a particular version of colonial nationalism which drew its energy and legitimacy from economic independence. The continuing importance of sheep and wool was noted fifty years later by Anthony Trollope in his *Australia and New Zealand*: "The squatter produces wool, and knows that wool is the staple produce of the colonies. To his thinking, success in wool means Australian greatness."[8] In 1930, a century after Cunningham, W.K. Hancock claimed even more dramatically that the industry "was Australia's opportunity, the grand occasion offered to the most wretched of colonies to pay its ransom and win its way to freedom and self-respect. Wool made Australia a solvent nation, and in the end a free one."[9]

---

[7] Quoted and discussed in Sidney Rosenberg, "Black Sheep and Golden Fleece: A Study of Nineteenth Century English Attitudes toward Australian Colonials" (unpublished doctoral dissertation, Columbia University, 1954): 43.

[8] Quoted in P.D. Edwards, Introduction to *Harry Heathcote of Gangoil*, by Anthony Trollope, ed. Edwards (Worlds Classics; Oxford: Oxford UP, 1992): xvi–xvii.

[9] Quoted in Stephen Dovers, "The History of Natural Resource Use in Rural Australia: Practicalities and Ideologies," in *Agriculture, Environment and Society: Contemporary Issues*

What is emphasized in the work of both earlier and later writers about the wool industry is that the fascination with sheep reflected the widely and fervently held belief that fortunes could be made from them. In 1836 Charles Darwin could comment regretfully that "the whole population, poor and rich, are bent on acquiring wealth: amongst the higher orders, wool and sheep-grazing form the constant subject of conversation."[10] (Trollope, on the other hand, claims to have found himself entirely engrossed by sheep, or at least by conversations about them.[11]) The interest discerned by these visitors reflects the fact that, while the first commercial exports of wool were made from the colonies to England in around 1806–07, by 1830 two million pounds were being exported, rising to ten million pounds by 1839, and more than double that amount by 1845.[12] While these figures need to be read in the context of a British domestic production of around 100 million pounds in the same period, the Australian exporters did make significant inroads in British markets.[13]

What was necessary to the levels of increase signalled here – which, unlike later ones, are attributable to rising numbers of stock rather than increases in carrying capacity and / or improved yield of wool per head – was the appropriation of land and the destruction of the lives and cultures of the Aborigines who shaped the very environment that had seduced the pastoralists. But as fire was used less and less, thick scrub began to overtake what had once been relatively open grazing lands. This in turn necessitated a major effort in clearing trees; thus Geoffrey Bolton suggests that from about the 1860s, "seeking clear grazing for their sheep," the pastoralists "went in for ring-barking on an enormous scale."[14] Bolton's comment implies a search: in fact, the need for and practice of clearing were *effects* of major changes to vegetation caused by pastoralism. Thus he goes on to note that "in districts such as the Lachlan and Mudgee after the development which followed the

---

for Australia, ed. Geoffrey Lawrence, Frank Vanclay & Brian Furze (Melbourne: Macmillan, 1992): 5.

[10] Quoted in Rosenberg, "Black Sheep and Golden Fleece," 56.

[11] In *Australia and New Zealand*. See Edwards, Introduction to *Harry Heathcote of Gangoil*.

[12] Rosenberg, "Black Sheep and Golden Fleece," 56, citing R.M. Crawford, *Australia* (London: Hutchinson, 1952): 61.

[13] Noel Butlin, *Forming a Colonial Economy: Australia 1810–1850* (Cambridge: Cambridge UP, 1994): 178.

[14] Bolton, *Spoils and Spoilers*, 42.

1861 *Land Act* it was found that the clearing of eucalypts was followed by a takeover of pine scrub" (43). C.E.W. Bean describes an entire pine forest formed by the action of sheep who trod down the soil and destroyed native vegetation – a forest subsequently killed by drought.[15]

Colonial and historical narratives about sheep must manage this environmental destruction, and they do it by attributing environmental transformation entirely to the sheep, rather than to the agency of the colonists who bring them, or the Aborigines who cleared the land into which they were brought; the main strategy for this management, in other words, is silence about human beings. There is a persistent effacing of the acts of dispossession that *are* "opening up the country" – the pastoral invasion which, ironically, produces the need for "clearing the land." A surprisingly recent example of this kind of erasure is the entry on the "Shearer in Australian Literature" in the *Oxford Companion to Australian Literature* (1994), which comments that "with the opening of new pastoral areas, central and western NSW, the Hunter Valley, the Darling Downs in Queensland, the Australia Felix districts of Victoria, thirty million hectares of grazing land became available for sheep farming."[16] It is this euphemization of invasion – "becoming available" – that magically substitutes pastoralist for indigene and legitimates pastoralism itself, allowing the new nation to ride, relaxed and comfortable, on the broad back of the sheep. The shift in the meaning of the term 'squatter' precisely marks this transformation: the word was first used to describe a man [sic] who, having acquired stock by dubious means, moved his sheep or cattle beyond the legal boundaries of white settlement, perhaps reappearing later as a person of substance; by the middle of the nineteenth century, the term denoted a person of entirely respectable means.

Part of the power of writers like Judith Wright, Eric Rolls and other commentators such as William Lines is that they lament the damage done by greed and ignorance in the nineteenth century, an appealing argument to an audience anxious to displace responsibility for environmental destruction onto earlier generations. It is widely believed that the pastoralists were oblivious to the damage done by their stock to the land and its vegetation, although in fact, when Griffiths asserts that the grasslands were lost in a generation, he understates the case:

---

15 C.E.W. Bean, *On the Wool Track* (London: Alston Rivers, 1910).

16 William H. Wilde, Joy Hooton & Barry Andrews, *Oxford Companion to Australian Literature* (Oxford: Oxford UP, 2nd ed. 1994): 691.

> In parts the sand that covers [Australia], and that holds the whole calendar of priceless seeds that have taken a few million years, at an underestimate, to evolve, is not more than one foot thick; so thin and light and delicate a skin that only the delicate Western scrub which grows from it holds it in its place at all. In certain parts, where men came out on to it and cut the scrub down recklessly, with rough-shod, ready-made European methods, the surface of the earth has blown clean away. All the exquisite wonderful plant-life of all the ages took *just one bad season to destroy*; and great patches of "scalded" clay stand to-day exactly as bare and with as hopeless a task to face as on the day when the last wavelet of a receding ocean lapped over them and left them to evolve some covering to their nakedness.[17]

This lyrical account from Bean is confirmed by later commentary on environmental history, in which it is pointed out that levels of stocking and rainfall are the crucial variables in terms of the different effects of occupation by sheep. Whereas environmental commentaries to date describe ungulate irruption – the rapid overpopulation of an area by hard-hoofed quadrupeds such as sheep or deer – as a species-specific response to initially hospitable environments, the Australian experience strongly suggests that irruption is *managed* overpopulation which occurs as colonists deliberately push to the limit the carrying capacity of newly colonized land.[18]

When popular historical writers like C.E.W. Bean address the impact of sheep on the environment directly – an impact that actually *constituted* colonization – we can, perhaps unexpectedly, find the celebration we so often find in pastoral histories very much tempered by a sense of great loss:

> Except for the sheep, and the sheep alone, the West would be, and is, to-day as the untamed centuries had left it – as the first white man, when he came over the red sandhill on the horizon on to the edge of the pine scrub, found it. Some way out of Menindie we happened to drive through a paddock that had been unstocked for many years. It gave one glimpse of what the white man did find. It was almost impossible to get out of one's head the idea that one was driving through a park.

Bean is told by an informant, who has lived for decades in the sheep country Bean is travelling through, that

---

17 Bean, *On the Wool Track*, 88 (emphasis added).

18 See, for example, Elinor C.K. Melville, *A Plague of Sheep: Environmental Consequences of the Conquest of Mexico* (Cambridge: Cambridge UP, 1994): 45–55.

> When first I rode my horse on to this red country, he sank into it up to his fetlocks at every step. It was all beautifully-grassed open country away to the hills – not a pine tree on it. But the soil was so loose that the sheep drove their feet deep into it as they walked over it.
>
> It was open country then. In about a year or two the sheep had trodden in the face of it and hardened it. [...] no sooner was the ground hard than up came this pine scrub thick all over the face of it.[19]

"Open country" and "a park" give no hint of human or indeed of any animal occupation: the land lies available to pastoralism. This rhetoric leaves out the problematical presence of the indigene, despite the fact that Aborigines were to prove crucial as a labour force in the sheep and cattle industries, a centrality that is figured repeatedly in fiction and memoir but was until the 1980s ignored by critics and historians.[20] Likewise, the "settlers" are vanished; the only occupants of this perfect place are plants.

The writer of these comments helped to make his name and reputation with them: C.E.W. Bean would later become Australia's first official war correspondent, and subsequently the official historian of the First World War. They come from *On the Wool Track*, made up of essays commissioned in 1909 for the *Sydney Morning Herald*, and published as a collection the following year.[21] *On the Wool Track* was popular enough to be reprinted three months later and again in 1912, with further editions in 1913, 1916, 1925, 1927, 1945, 1947, 1963, 1967, 1969 and 1985. Bean's book brings together representations of sheep, environment and invasion in ways that both typify and call into question the arguments being made in current debates over environmental and colonial history.

In two chapters in particular, "The Red Country" and "What Happened," Bean eulogizes the country, specifically the soil and vegetation lost to sheep. He "quotes" the stock inspector of the Cobar sheep district:

> "In the years 1880 and 1881, before this district was stocked, and when it was being improved [...] the country was covered with heavy growth of natural grasses – kangaroo grass, star grass, blue grass, mulga, and other grasses. The western half of the district abounded with salt and cotton bush, together with the grasses mentioned. The ground was soft, spongy, and very absorbent. One

---

19 Bean, *On the Wool Track*, 5, 9.

20 See, for example, Dawn May, *Aboriginal Labour and the Cattle Industry: Queensland from White Settlement to the Present* (Melbourne: Cambridge UP, 1994).

21 Wilde, Hooton & Andrews, *Oxford Companion to Australian Literature*, 85.

> inch of rain, then, in spring or autumn, produced a luxurious growth of fresh green grass." (94)

Bean argues that "If anything ever needed tender, scientific handling, it was this tender covering of grass and trees" (93). The land is mourned but at this key moment its invaders, like its first keepers, are forgotten, as the sheep and the sheep alone are held responsible for the invasion and destruction of the land. It also erases the actual *contest* for the land that Henry Reynolds has called "the Black Wars," the open and covert conflict which he estimates killed some 20,000 people during colonization.[22]

The writing of Reynolds has perhaps been the most important catalyst for contemporary debate in academia on relations between pastoralists and Aborigines, his study *The Law of the Land* being central to the rethinking of the *terra nullius* legal precedent that enabled the *Mabo* decision. Reynolds interrogates the commonly held assumption that the Aborigines were held *not* to be in possession of land because they did not cultivate it, pointing out that the idea of cultivation being necessary to prove possession was not merely a cultural fiction, but a legal one as well. More importantly, the argument that colonization was entirely legitimate because Aborigines did not farm the land was, as Reynolds suggests, "a strange argument to advance in a pastoral country where only a tiny proportion of the land was under crop."[23] Reynolds's basic argument is that "the common law was corrupted in Australia by the nature of the relationship between settlers and Aborigines in the same way in which it was corrupted in Britain's slave colonies," adapting itself to accommodate dispossession.[24]

In his commentary "Land Rights, Then and Now," Henry Reynolds asserts that a majority of the colonists "were willing to pay the cost of forced expropriation" – illegal seizure of land – "in insecurity, property loss, suppressed guilt. They could bear all that because the real price was paid in the destruction and desolation of Aboriginal tribes all over the continent."[25] But there are few signs of insecurity, loss or guilt in pastoral histories. Typical is

---

[22] Henry Reynolds, *Frontier: Aborigines, Settlers and the Land* (Sydney: Allen & Unwin, 1987): 189.

[23] Henry Reynolds, *The Law of the Land* (Ringwood, Victoria: Penguin, 1987): 75.

[24] Reynolds, *The Law of the Land*, 4. It is a parallel hinted at in Rolf Boldrewood's discussion of colonial architecture spreading from the southern states of the USA and the West Indies in *A Colonial Reformer* (London: Macmillan, 1895): 39.

[25] Reynolds, *The Law of the Land*, 154.

Bean, who touches on the question in just one paragraph of *On the Wool Track*:

> There have been men, no doubt, living on the fringe of settlement whose deeds will never be told if only because those who knew them would not have the courage to tell them. One very old pioneer on the Darling said to us: "I sometimes think the drought came upon the squatters as a retribution for some of the wickedness that has been done."[26]

These men are inside yet outside settlement, "on the fringe"; outside yet inside narrative – their "deeds will never be told" and are mentioned here only as being unmentionable. These are stories that people fear to tell, although they are remembered by "one very old pioneer." The paragraph precisely enacts the "great forgetting" of Australian history identified by W.E.H. Stanner in his influential 1968 ABC Boyer Lectures.[27]

The agent of invasion in this narrative is not the soldier, the explorer, the pastoralist or his shepherd: it is the sheep. The animal stands in as colonial "occupant" of the land. This was literally and legally the case: land found to be unoccupied by stock could be allocated to another, hence the scramble for stock in the 1820s and 1830s. That sheep were a particularly strong guarantee of possession was signalled legislatively in the passing, in 1843, of the *Liens on Wool and Mortgages on Stock Act*, which made sheep and their wool valid security for loans. This was followed by an Order in Council of 1847 which confirmed squatters' occupation of lands beyond authorized limits of (white) settlement.[28] In its extreme form, this rhetoric of displacement makes the sheep the invader; it is the animals themselves that take over the land and literally expunge the traces of Aborigines

> Whatever the sheep may or may not have done, they have done this for Australia. As far as the West is livable for men, it is the sheep and sheep alone that have made it so [...] Where the Australian country has driven back the sheep [...] it has driven the white man too [...] and the land is desolate, fences down, homesteads ruined.[29]

---

26 Bean, *On the Wool Track*, 47.

27 W.E.H. Stanner, *After the Dreaming: Black and White Australians: An Anthropologist's View* (Sydney: Australian Broadcasting Corporation, 1969).

28 Alan Barnard, *The Australian Wool Market 1840–1900* (Melbourne: Melbourne UP, on behalf of the Australian National UP, 1958): 9.

29 Bean, *On the Wool Track*, 11, 15.

The race is repelled, its boundaries broken, its domestic centre destroyed, their success having been and continuing to be contingent on the colonizing energies of the sheep.

There is a curious echo of Bean's work in the writing of farmer and novelist Eric Rolls, one of the more popular practitioners of the kind of revisionist geography and botany that figures colonization as environmental destruction. Rolls' seductive prose tends to deploy a quasi-maternal set of metaphors focused on nurturing and neglect that casts colonial culture not as rapacious capitalist, but rather as negligent parent:

> When Europeans came to Australia, the soil had a mulch of thousands of years. The surface was so loose you could rake it through the fingers. No wheel had marked it, no leather heel, no cloven hoof. Digging sticks had prodded it, but no steel shovel had ever turned a full sod. Our big animals did not make trails. Hopping kangaroos moved in scattered company, not in damaging single file like sheep and cattle. [...] Every grass-eating mammal had two sets of sharp teeth to make a clean bite. No other land had been treated so gently.[30]

This passage, from his compellingly titled essay "More a New Planet than a New Continent," disembodies the indigenous inhabitants, whose presence is marked only by the "prodding" of the digging sticks. So, too, are the colonizers erased: they are wheels, leather heels, unnamed and unimagined as the land itself is foregrounded. In precisely the same rhetorical manoeuvre as that effected by Bean, the materiality of colonization is displaced onto animals.

This manoeuvre, in which animals become empire's proxy in the work of colonization, is by no means limited to Australia. One station owner in New Zealand commented of his flock that

> The sheep of the station are, in sober truth, working out their own solution. They are returfing the naked windblows, hardening the erstwhile dangerous fords, drying the bogs and marshes, building viaducts, shaping sleeping-shelves, exposing pitfalls and chasms. They are remodelling the run to suit their particular requirements. In this good work of reclamation other stock have participated. It is the sheep, however, that has borne the burden and the heat of the day: it is owing to him that for his race the run is more easy to perambulate, more safe to traverse.[31]

---

[30] Eric Rolls, "More a New Planet than a New Continent," in Dovers, ed. *Australian Environmental History*, 22.

[31] Herbert Guthrie–Smith, *Tutira: The Story of a New Zealand Sheep Station* (1921; Wellington: A.H. & A.W. Reed, 4th ed. 1969): 180.

The language of these comments alerts us to the colonial impetus of the narrative: the sheep is saving the land for his 'race' (not breed as we might expect), drying out those 'bogs and marshes' that will entrap the unwary, exposing 'pitfalls and chasms' that will likewise entrap those who traverse the country. The goal here is safety, certainty and, ultimately, productivity.

Likewise, in South Africa the techniques of land management and exploitation that characterized Australian pastoralism – techniques which hastened the destruction of indigenous communities – were copied and deployed in order to accelerate the dismantling of key elements of indigenous life-styles. "The imperatives of pastoral reform became intertwined with concerns about population growth, urban migration, social control and the maintenance of segregation" in a community of farmers, veterinarians and administrators who "were aware of innovation in Britain and in other settler colonies, particularly Australia."[32] African practices of livestock management, particularly kraaling, were seen not only as contributing to disease but were also perceived to be implicitly immoral[33] – control of vegetation, soils, disease, herding and breeding of stock were grounded in land and stock management practices copied from Australia that deliberately accelerated the takeover of indigenous land and the dispersal of indigenous communities.

Physical colonization of the Australian continent, and the transformation of stock management practices in South Africa and New Zealand in accordance with the model established in Australia, was done not by soldiers nor explorers nor missionaries, but by those aiming to find and seize land on which they could pasture sheep and cattle. The narratives that survive from *these* first travels are much simpler ones than explorers' journals or official histories: they are claims for land, and orders for supplies. Henshaw Jackson commented in 1910 that "Someone has said that the bond of Empire is a silken thread [...] but there is no denying the fact that the threads which connect the far corners of the earth with Australia are made of wool."[34] It is for this reason that the songs of praise to the sheep are not composed in the scientific and agricultural discourses we might expect, but a distinctive mix of the commercial, the environmental and the imperial. The sheer produc-

[32] William Beinart, "Vets, Viruses and Environmentalism at the Cape," in *Ecology and Empire: Environmental History of Settler Societies*, ed. Tom Griffiths & Libby Robin (Melbourne: Melbourne UP, 1997): 87, 88.

[33] Beinart, "Vets, Viruses and Environmentalism at the Cape," 93.

[34] Henshaw Jackson, *Broken Fleece* (Sydney: William Brooks, 1910): 107.

tivity of the sheep intoxicates the capitalist as quickly and completely as it horrifies those who observe the effects of the animal on soils and vegetation. In the absence of the British race, specifically the yeoman farmer so long and so often sought by the authorities to fill the "empty spaces" of the Australian landmass, it is the white-woolled sheep, the "pure merino," that will populate – and devastate – the interior of the country, for the enrichment of empire and, later, a largely urban-generated nationalist mythology.

WORKS CITED

Barnard, Alan. *The Australian Wool Market 1840–1900* (Melbourne: Melbourne UP, on behalf of the Australian National UP, 1958).

Bean, C.E.W. *On the Wool Track* (London: Alston Rivers, 1910).

Beinart, William. "Vets, Viruses and Environmentalism at the Cape," in Griffiths & Robin, ed. *Ecology and Empire*, 87–101.

Boldrewood, Rolf. *A Colonial Reformer* (London: Macmillan, 1895).

Bolton, Geoffrey. *Spoils and Spoilers: A History of Australians Shaping Their Environment* (Australian Experience series; Sydney: Allen and Unwin, 2nd ed. 1992).

Brunet, Sandra. "Landscape and Identity: Judith Wright's Fragile Land and Haunted Self' (unpublished MA thesis., University of Queensland, 1998).

Butlin, Noel. *Forming a Colonial Economy: Australia 1810–1850* (Cambridge: Cambridge UP, 1994).

Cary, John, & Neil Barr. "The Semantics of 'Forest Cover': How Green was Australia," in Lawrence, Vanclay & Furze, ed. *Agriculture, Environment and Society*, 60–76.

Dovers, Stephen. "The History of Natural Resource Use in Rural Australia: Practicalities and Ideologies," in Lawrence, Vanclay & Furze, ed. *Agriculture, Environment and Society*, 1–18.

Dovers, Stephen, ed. *Australian Environmental History: Essays and Cases* (Oxford: Oxford UP, 1994).

Edwards, P.D. Introduction to *Harry Heathcote of Gangoil*, by Anthony Trollope, ed. Edwards (Worlds Classics; Oxford: Oxford UP, 1992): vi–xviii.

Flannery, Timothy. *The Future-Eaters: An Ecological History of the Australasian Lands and Peoples* (Port Melbourne: Reed, 1994).

Frawley, Kevin. "Evolving Visions: Environmental Management and Nature Conservation in Australia," in Dovers, ed. *Australian Environmental History*, 55–78.

Griffiths, Tom, & Libby Robin, ed. *Ecology and Empire: Environmental History of Settler Societies* (Melbourne: Melbourne UP, 1997).

Guthrie–Smith, Herbert. *Tutira: The Story of a New Zealand Sheep Station* (1921; Wellington: A.H. & A.W. Reed, 4th ed. 1969).

Hallam, Sylvia. "The First Western Australians," in *A New History of Western Australia*, ed. C.T. Stannage (Perth: U of Western Australia P, 1981): 35–71.

Jackson, Henshaw. *Broken Fleece* (Sydney: William Brooks, 1910).

Lawrence, Geoffrey, Frank Vanclay & Brian Furze, ed. *Agriculture, Environment and Society: Contemporary Issues for Australia* (Melbourne: Macmillan, 1992).

Lines, William. *Taming the Great South Land: A History of the Conquest of Nature in Australia* (Sydney: Allen & Unwin, 1991).

May, Dawn. *Aboriginal Labour and the Cattle Industry: Queensland from White Settlement to the Present* (Melbourne: Cambridge UP, 1994).

Melville, Elinor G.K. *A Plague of Sheep: Environmental Consequences of the Conquest of Mexico* (Cambridge: Cambridge UP, 1994).

Reynolds, Henry. *The Law of the Land* (Ringwood, Victoria: Penguin, 1987).

——. *Frontier: Aborigines, Settlers and the Land* (Sydney: Allen & Unwin, 1987).

Rolls, Eric. "More a New Planet than a New Continent," in Dovers, ed. *Australian Environmental History*, 22–36.

Rosenberg, Sidney. "Black Sheep and Golden Fleece: A Study of Nineteenth Century English Attitudes toward Australian Colonials" (unpublished doctoral dissertation, Columbia University, 1954).

Stanner, W.E.H. *After the Dreaming: Black and White Australians: An Anthropologist's View* (Sydney: Australian Broadcasting Corporation, 1969).

Wilde, William H., Joy Hooton & Barry Andrews. *The Oxford Companion to Australian Literature* (Oxford: Oxford UP, 2nd ed. 1994).

Young, Ann R.M. *Environmental Change in Australia since 1788* (Melbourne: Oxford UP, 1996).

Representations of Landscape and Nature in Anthony Trollope's *The West Indies and the Spanish Main* and James Anthony Froude's *The English in the West Indies*

CLAUDIA BRANDENSTEIN

IN HER INFLUENTIAL WORK *Imperial Eyes*, Mary Louise Pratt demonstrates that an important convention in imperial travel and exploration narratives is the production of the monarch-of-all-I-survey scene. This frequently entailed climbing up a mountain, taking in and taking down, from "some noble coign of vantage,"[1] the panorama that lay before "the eye of the beholder." Pratt gives the example of Richard Burton's discovery of Lake Tanganyika, pointing out that the

> 'discovery' of such sites/sights involved making one's way to the region and asking the local inhabitants if they knew of any big lakes, etc. in the area, then hiring them to take you there, whereupon with their guidance and support, you proceeded to discover what they already knew.[2]

This is, according to Pratt, a crucial way in which imperialists take possession. Another characteristic of this rhetorical mode is the elision of the network of

---

[1] This phrase was used by Henry Morton Stanley, who journeyed to the African interior in search of David Livingstone.

[2] Mary Louise Pratt, *Imperial Eyes: Travel Writing and Transculturation* (London: Routledge, 1992): 202. Further page references are in the main text.

assistants, in the shape of local inhabitants, which makes these ventures possible in the first place. I want to begin this reading of the representation of nature and landscape in Anthony Trollope's *The West Indies and the Spanish Main* and James Anthony Froude's *The English in the West Indies*[3] by considering a West Indian version of the monarch-of-all-I-survey motif found in Trollope's travelogue.

In 1858, Trollope was selected for the job of reorganizing the postal system in the West Indies. On 1 November, he was in London making preparations for the journey and by 17 November was outward bound for the Spanish Main. On first encountering the mountains of Jamaica, the writer declares, "There is scenery in Jamaica which almost equals that of Switzerland and the Tyrol."[4] Having been thus inspired by the beauty and grandeur of these mountains, which are magnified because of their near likeness to European landscape, Trollope explains that he will organize an expedition to Jamaica's Blue Mountain Peak. He describes in rapturous detail the "grandeur of scenery" and "lovely park-like landscapes" at the foot and lower regions of the mountains and, after dwelling at some length upon the trials and tribulations of the journey, he confesses that his earlier, somewhat lofty, ambitions have been thwarted, his tone changing as he begins the ascent in earnest:

> And now I have a mournful story to tell. Did any man ever know of any good befalling him from going up a mountain? [...] as for the true ascent – the nasty, damp, dirty, slippery, boot-destroying, shin-breaking, veritable mountain! Let me recommend my friends to let it alone [...] I have tried many a mountain in a small way, and never found one to answer. I hereby protest that I will never try another. (38)

---

[3] After Trollope had returned home to England, a phrase was coined to describe his visit to the West Indies – "to go a-Trolloping"; this phrase referred to the way in which the novelist hurried from island to island forming hasty conclusions along the way. The local language was also enriched by the title of J . J . Thomas' response to the historian J . A . Froude's island-hopping expedition – *Froudacity*. Froude left England in late December 1887 and was home again by April of the following year. Froude's book *Oceana* (1886), about his brief tour in Australasia, caused much offence among colonists; someone in Australia coined the term "froudacious" to describe this travelogue.

[4] Anthony Trollope, *The West Indies and the Spanish Main* (1859; Gloucester: Alan Sutton, 1985): 20. Further page references are in the main text.

Trollope's description of his preparations to ascend the Blue Mountain Peak and his subsequent account of the ascent is unconventional, parodying as it does the discourse of conquest, exploration and discovery in a number of ways. From the outset he adopts a satirical tone, stating, "we were to spend the night on the Blue Mountain Peak, in order that the rising sun might be rightly worshipped" (38). On his ascent, Trollope and his party meet, as he puts it, a "hospitable coffee-planter" whose "tales as to the fate of other travellers made me tremble for what might some day be told of my own adventures" (38–39). It is significant, too, that Trollope points to the elaborate process of planning and organization that enables the journey: "I provided myself with a companion, and he provided me with five negroes, supply of beef, bread, and water, some wine and brandy, and what appeared to me to be about ten gallons of rum" (38). He acknowledges the source of the labour, that is, the black locals who make the journey at all possible; in his account they become visibly present. Food and other provisions do not appear out of nowhere: "The five negroes each had loads on their heads and cutlasses in their hands. We ourselves travelled without other burdens than our own big sticks" (39).

"The verbal painter," as Pratt has noted, "must render momentously significant what is, practically, a non-event" (202). Rather than constructing his mountain-climbing expedition as "momentously significant," Trollope deliberately figures his excursion as a non-event, pointing to the lack of adventure. The absence of action thus becomes the narrative's focal point:

> I have nothing remarkable to tell of the ascent. We soon got into a cloud, and never got out of it. But that is a matter of course. We were soon wet through up to our middles, but that is a matter of course also. We came to various dreadful passages, which broke our toes and our nails and our hats, the worst of which was called Jacob's ladder – also a matter of course [...] We were wet through and through, and could hardly see twenty yards before us on any side. (39)

Thus Trollope cannot, as was common in the monarch-of-all-I-survey mode, evaluate or possess this scene. From this lofty height the author can see nothing, the very point of the narrative becoming the spectacular failure of the expedition:

> Slowly and mournfully we dried ourselves at the fire; or rather did not dry ourselves, but scorched our clothes and burnt our boots in a vain endeavour to do so. It is a singular fact, but one which experience has fully taught me, that when a man is thoroughly wet he may burn his trousers off his legs and his shoes

> off his feet, and yet they will not be dry – nor will he. Mournfully we turned ourselves before the fire – slowly, like badly-roasted joints of meat; and the result was exactly that: we were badly roasted – roasted and raw at the same time. (40)

Trollope turns his journey to the mountain peak into comedy and a drunken farce, and thus deflates the hyperbole of imperial travel writing. By refusing to participate in the discourse of colonization and conquest, and by declining to take himself seriously, the writer creates an anti-(imperial)hero, a bumbling adventurer, who cannot conquer and tame the obstacles he encounters; instead, he depicts himself as having been conquered.

A similar response to panoramic land scanning is at work in James Anthony Froude's travel narrative. During his sojourn in Dominica, the author explains that "A very particular object was to reach the crest of the mountain ridge which divides Dominica down the middle. We saw the peaks high above us, but it was useless to try the ascent if one could see nothing when one arrived."[5] He continues: "To ride up a mountain three thousand feet high, to be near a wonder which I could not see after all, was not what I had proposed to myself" (149). The reader is told that "panoramic views from mountain tops, extolled as they may be, do not particularly interest me" (202). The writer makes quite clear that he has no interest in adventuring nor in producing an adventure narrative. Froude emphasizes that he has more important business to attend to than climbing mountains in quest of panoramic views; for Froude there is no value in this kind of relationship with landscape. I want to consider why there is here an almost deliberate rejection of a convention which is such a fundamental feature of nineteenth-century male exploration and travel narrative. Froude and Trollope's responses become less puzzling when we take into account Chris Tiffin's observation that "imperial teleology had a work-ethic whose watchwords were 'use,' 'function,' 'progress'."[6] Indeed, the key to Anthony Trollope's response to mountain scenery is revealed later in his travelogue on his visit to Demerara:

> For those who wander abroad in quest of mountain scenery it must be admitted that this colony has not much attraction [...] go where you will, travel thereabouts as far as you may, the eye meets no rising ground. Everything stands on the same

[5] James Anthony Froude, *The English in the West Indies* (London: Longmans, 1888): 148. Further page references are in the main text.

[6] Chris Tiffin, "Progress and Ambivalence in the Colonial Novel," in *Re-Siting Queen's English: Text and Tradition in Post-Colonial Literatures*, ed. Gillian Whitlock & Helen Tiffin (Cross / Cultures 7; Amsterdam & Atlanta GA: Rodopi, 1992): 3.

> level. But then, what is the *use* of mountains? You can grow no sugar on them, even with ever so many Coolies. They are big, brown, valueless things, cumbering the face of the creation; very well for autumn idlers when they get to Switzerland, but utterly useless in a colony which has to count its prosperity by the number of its hogsheads. Jamaica has mountains, and look at Jamaica! (130; my emphasis)

A different mode of subjugation is needed in a land where flat and fertile plains are at a premium and are crucial to the imperial economy; there is no need for heroics, or room for aesthetic indulgence, just hard work. This landscape, described by Trollope, is decidedly feminized, bereft of phallic objects and hence, it is suggested, easily subdued. Mountains, by contrast, prove to be intractable; they are thought of, in the nineteenth-century context of the West Indies, as obstacles to progress and are thus regarded as highly vexing to the imperial spirit. They are, therefore, no competition for plantation fields or the cultivated garden – those scenes which make imperial hearts beat faster. Froude and Trollope would, it seems, prefer to see the mountains levelled.[7]

For Anthony Trollope,

> Barbados is a very respectable little island, [which] makes a great deal of sugar. It is not picturesquely beautiful, as are almost all the other Antilles, and therefore has but few attractions for strangers. But this very absence of scenic beauty has saved it from the fate of its neighbours. A country that is broken into landscapes, that boasts of its mountains, woods, and waterfalls, that is regarded for its wild

[7] Interestingly, such responses to mountains and the uncultivated natural landscape are largely absent in travel narratives written at the height of sugar production in the West Indies, such as *Lady Nugent's Journal* and "Monk" Lewis's *Journal of a Residence Among the Negroes*. Lewis states that the north side of the island of Jamaica is "extremely beautiful and sublime" and that "the beauty of the atmosphere [and] the dark purple mountains" produced "a very picturesque appearance"; Matthew Gregory Lewis, *Journal of a Residence Among the Negroes in the West Indies* (London: John Murray, 1845): 22–24. In her journal, Lady Nugent describes a scene on the Jamaican bog walk as "most romantic, beautiful and picturesque" and refers to "the roar of the river beneath, which was quite sublime"; Maria Nugent, *Lady Nugent's Journal* (Kingston, Jamaica: Institute of Jamaica, 1966): 61. In these works, in which landscape is domesticated according to principles of the picturesque and the sublime, uncultivated sites are perceived as enhancing the environment and consequently do not give cause for disquiet. A quite different response to and politics of landscape depiction is in operation in these earlier writings.

> loveliness, is seldom propitious to agriculture. A portion of the surface in all such regions defies the improving farmer. (153)

According to the writer, "It is the waste land of the world that makes it picturesque. But there is not a rood of waste land in Barbados" (154). This is in stark contrast to his description of Jamaica, a once prosperous and highly lucrative part of the British empire which, in Trollope's terms, has become "one of the few sores in [the British empire's] huge and healthy carcase" (78). Trollope thus turns his attention to other uses to which this island may be put; "all Jamaica", he says, "is suited for a grazing-ground, and all the West Indies should be market for their cattle" (30).

In *A Small Place*, Jamaica Kincaid writes: "we, for as long as we have known you, *were* capital, like bales of cotton and sacks of sugar."[8] Kincaid is, of course, referring to the imperialist and neo-imperialist exploitation of African slaves and their descendants in the West Indies, but her comment is equally suggestive of the ways in which the islands of the West Indies have been imagined or considered as resources for European exploitation – treasure islands to fill the coffers of avaricious imperialists – since Columbus' encounter with the new world. The West Indian environment is perceived first and foremost as capital to be exploited. A number of sometimes conflicting master-discourses are put into play in the narratives under consideration here, particularly that of the New World as a paradisal garden of Eden. Froude especially produces the West Indies as an abundant paradise, an Eden awaiting rediscovery, containing inexhaustible *and* unimagined resources.

In considering the modes of response to landscape generated in the narratives, it is necessary to consider the historical context in which these travelogues operate. Both Trollope's *The West Indies and the Spanish Main* and Froude's *The English in the West Indies* were written after the decline of the sugar industry in the Caribbean, when it had ceased to be an economically viable enterprise.[9] Britain, at the time Froude's text was written in 1888, had its imperial sights trained on Africa; the scramble for Africa was taking place at the time of its composition. His book appears at a moment when the power of the plantocracy was reaching the final stage of its decline and experiments with representative government and the first steps towards

---

[8] Jamaica Kincaid, *A Small Place* (New York: Plume, 1988): 37.

[9] Debates concerning the fate of the islands had been taking place since the abolition of the slave trade and emancipation of slaves were first considered.

universal suffrage were being made. Froude, calling himself "an Englishman proud of his country," attempts to revive British interest in the West Indies. His argument is that "The West Indies are but a small limb in the great body corporate of the British Empire, but there is no great and no small in the life of nations. The avoidable decay of the smallest member is an injury to the whole" (320).[10] The writer emphasizes that there is an urgent need to heal this ailing body part. *The English in the West Indies* coincides with the period of high imperialism when Britain's anxiety over its colonies was at its height; Froude's fear is that "the great empire on which the sun never sets will be shattered to atoms" (184). He warns his home audience that "it is to America [Britain's rival in the region] that [...] trade is drifting" (97) and thus urges the continuing maintenance of the boundaries of the British Empire. A profound sense of anxiety, insecurity and desperation haunts this work. Trollope's *The West Indies and the Spanish Main*, written mid-century, while engaging in the debate over what to do with the West Indies and sharing many of the views propounded in Froude's later work, is much less angst-ridden and panic-stricken. As Patrick Brantlinger has noted, "the early and mid-Victorians [...] did not feel self-conscious or anxious about their world domination."[11]

Since Froude's text is plagued by an extreme sense of insecurity regarding the fate of the Empire, his investigation of other uses for the West Indies takes on rather different and larger proportions than that of Trollope. Putting out an urgent call for British settlers and speculators, the author travels as a scout for British investors, in the service of the greater good; his narrative functions as a report home of the – as yet unrealized – wonders of the new world. In significant ways, he rehearses the tropes of the original encounter with the Americas, particularly in terms of the promise he makes to his home audience that riches abound in the islands.[12] Ever on the lookout for exploit-

---

10 This metaphor of empire as a body is one which is repeated throughout the book.

11 Patrick Brantlinger, *Rule of Darkness: British Literature and Imperialism, 1830–1914* (Ithaca NY: Cornell UP, 1988): 23.

12 Froude's travelogue echoes, in striking ways, a letter written by Columbus summarizing his first trip to the New World, in which he describes the "natives" as "trustworthy and most liberal with all that they have; none of them denies to the asker anything that he possesses; on the contrary they themselves invite us to ask for it [...] [they] always give much for little, content with very little or nothing in return"; Marvin Lunenfeld, *1492: Discovery, Invasion, Encounter* (Sources in Modern History series; Lexington MA: D.C. Heath, 1991): 133. This figuration of the New World as beckoning or, rather, beseeching those from the Old World to take possession of what it has to offer is replayed in *The*

able resources, Froude includes information on potential ventures, labour conditions, transport and possible markets, as well as lengthy suggestions for improvement. Advocating the pursuit of new and different imperial dreams such as the cultivation and location of markets for oranges and tobacco, among other commodities, he points out that uncultivated and unused regions abound in these British possessions, and that land is once again going to waste. The author shifts the imperial gaze from sugar to other economic enterprises. Froude informs his audience that there is a need to retake possession and to re-imagine the possibilities of the West Indies, arguing that new triumphs and conquests are not only possible but are vital to the healthy functioning of the body politic. "Skill and capital and labour," his readers are advised, "have only to be brought to bear together, and the land might be a Garden of Eden. All precious fruits, and precious spices, and gums, and plants of rarest medicinal virtues will spring and grow and flourish for the asking" (140). Froude emphatically suggests that economic (ad)ventures of all kinds are still possible in the islands: "In the West Indies there is indefinite wealth waiting to be developed by intelligence and capital" (79). In significant ways, the writer *re-invents* the West Indies for British consumption.

Gazing upon an "uncultivated" or "unimproved" site in Dominica, Froude comments with regret, "here was all this profusion of nature, lavish beyond example, and the enterprising youth of England were neglecting a colony which might yield them wealth beyond the treasures of the old sugar planters" (141). Froude's pragmatically economic rhetoric makes overt what is in other West Indian travel texts much more implicit; this is not veiled or covert discourse but, rather, a quite blatant expression of imperial greed. The ultimate crime that one can commit against empire is, it seems, to let resources lie in waste, or to fail to realize their potential. Advocating an ethics of landscape use, the author emphatically suggests that Britons are under a moral obligation to use and make profit from the land. Casting his eye over cane fields which have been abandoned, Froude exclaims, "A state of things more helplessly provoking was never seen" (140). This land, he suggests, is crying out for redevelopment.

English in the West Indies. In this work, as in Columbus' writings, it is implied that the colonization of the region by Europe (more specifically, in this case, Britain) has been destined by divine forces and that to fail to heed this call is to fail to do one's duty to God and country.

Throughout Froude's text images appear of once cultivated sites which have been left to run wild. The trope that recurs with greatest frequency is that of the West Indies as a once cultivated garden which has relapsed into wilderness: "Lands under once high cultivation," the reader is informed, "are lapsing into jungle" (250). Nature has almost erased the signs of civilization or, more precisely, the impress of the imperial stamp, and now threatens to undo all the hard work that, the writer argues, the English have put into the making of the West Indies.[13] Froude advises that Britain should aspire to re-create in its West Indian possessions the achievement of Barbados, which he holds up (as does Trollope) as a monument to British accomplishment. Speaking of this island, he notes with pride that it was originally "covered with forest. In the interior little remains save cabbage palms and detached clumps of mangy-looking mahogany trees. The forest is gone, and human beings have taken the place of it" (104). He finds, while touring the island, that there is no sign of "neglect in the fields [...] there was not a weed to be seen or broken fence where fence was needed" (100). The author perceives, or constructs, the West Indies as a garden that requires careful tending and pruning, one that is in need of the ever-watchful presence of the British imperial eye / I. He cautions his readers on numerous occasions that the islands will revert to savagery and barbarism if not properly managed. The image of the cultivated garden returning to wilderness functions as a metaphor in his travelogue for the decline of British interest and the disintegration of British power and influence in the Caribbean.

Although Froude expresses misgivings concerning panoramas and mountain scenery, he does relent and eventually ascends a mountain peak in Dominica. When "a tolerable morning" for the ascent finally arrives, the author finds himself "3,000 feet above the sea," where

> Far away the ocean stretched out before us, the horizon line where sky met water so far distant that both had melted into mist at the point where they touched [...] Below, above, around us, it was forest everywhere; forest, and only forest, a land fertile as Adam's paradise, still waiting for the day when 'the barren woman shall bear children.' Of course it was beautiful, if that be of any consequence – mountain peaks and crags and falling waters, and the dark green of the trees in the foreground, dissolving from tint to tint to grey violet, and blue in the far-off distance. (151)

---

[13] In this, as in much of the large body of writings on empire, the labour of the colonized and the enslaved has been systematically erased.

A number of conventions of landscape representation are combined in this description. The scene, constructed according to the principles of the picturesque, is composed as a painting with foreground and background and a focus on variations in colour. The writer's eye moves out to the horizon and then returns to inspect and appraise his immediate surroundings. The picturesque is interlaced with the monarchic view and lamentations over land which is lying in waste. Froude frequently pauses to sketch; the engravings, included in his travelogue, are reproductions of sketches he made while touring the islands. These engravings, depicting managed and orderly picturesque sites where nature is neatly framed and contained, are at odds with the verbal pictures of the West Indies which the writer paints in his travelogue. To an extent, Froude fashions himself as a late-nineteenth-century version of the eighteenth-century traveller touring Europe in search of the picturesque. Froude, however, does not pause in order to appreciate the wild beauty of the "uncultivated" sites he encounters; rather, they become occasions to express his concern that there is no evidence of the improving hand of imperialism. Land which is unimproved, or incapable of improvement, has little appeal for the imperial observer. Indeed, this observer often turned away in revulsion. According to Mary Louise Pratt, in the travel accounts of South America written by Britons in the first few decades of the nineteenth century, "unexploited nature tends to be seen [...] as troubling or ugly, its very primal qualities a sign of the failure of human enterprise" (149). That this idea is at work in Froude's text is clear in the following passage, in which he scans a scene in Dominica from the crest of the mountain ridge which divides the island:

> What a land! And what were we doing with it? This fair inheritance, won by English hearts and hands for the use of the working men of England, and the English working men lying squalid in the grimy alleys of crowded towns, and the inheritance turned into a wilderness. (151)

Pratt notes that an important component of the production of the monarch-of-all-I-survey scene is the transformation of the land being scanned from wilderness to productive site. In attempting to participate in this discourse of conquest and discovery, Froude prepares to impose his vision of progress on the scene he is scanning and to imagine the civilizing mission in full swing. Ironically, however, just as visions begin "to rise of what might be" they are taken from Froude "before they could shape themselves. The curtain of vapour fell down over us [...] and all was gone and of that glorious picture nothing was left" (151). The landscape refuses his imperial musings and renders him sightless as he finds himself enveloped in mist. With his panoramic vision blinded, he cannot exert mastery over this scene and put Britain into the picture. The "curtain of vapour" puts an end to description and thus mutes the language of imperialism.

Imperial discourses are put to the test in one way or another throughout the narrative. One of the most immediate challenges to imperialism addressed in *The English in the West Indies* is from the descendants of Britain's slaves. I want to turn the discussion, at this point, to the representation of the impact of black West Indians on the Caribbean landscape in the texts under consideration. The tours being examined are, as has been suggested, to an extent predicated on the principles of picturesque travel, which involved the ordering of sites / sights according to specific aesthetic rules and codes so as to produce pleasing narrative effects. One object which, to Anthony Trollope's aesthetic sense, does not produce such a pleasing effect is a black West Indian village. Spatial order cannot be created out of such a site / sight for, according to the author, it defies the picturesque aesthetic frame. He describes the houses and cottages in a "negro village" as standing

> in extreme disorder, one here and another there, just as individual caprice may have placed them. There seems to have been no attempt at streets or lines of buildings, and certainly not at regularity in building. Then there are no roads, and hardly a path to each habitation. As the ground is not drained, in wet weather the whole place is half drowned. Most of the inhabitants will probably have made some sort of dyke for the immediate preservation of their own dwellings; but as those dykes are not cut with any common purpose, they become little more than overflowing ponds, among which the negro children crawl and scrape in the mud; and are either drowned or escape drowning, as Providence may direct. The spaces between the buildings are covered with no verdure; they are mere mud patches, and are cracked in dry weather, wet, slippery, and filthy in the rainy season. (138–39)

In this representation, the black West Indian village appears primitive and crude, evincing little sign of civilization or cultivation. Trollope's description serves a dual ideological purpose: first, to suggest that black West Indians are in need of the guiding hand of mother England; and second, to erase all traces of an economy responsible for such apparent poverty. This contrasts with the pastoral images which characterize those travel narratives written during the era of plantation slavery, such as Monk Lewis' *Journal of a Residence Among the Negroes in the West Indies.* Lewis describes "the hermitage-like appearance of the negro buildings, all situated in little gardens and embosomed in sweet-smelling shrubberies"[14] and later states:

> I never witnessed on the stage a scene so picturesque as a negro village. I walked through my own to-day, and visited the houses of the drivers, and other principle [sic] persons; and if I were to decide according to my own taste, I should infinitely have preferred their habitations to my own. Each house is surrounded by a separate garden, and the whole village is intersected by lanes, bordered with all kinds of sweet-smelling and flowering plants; but no such gardens as those belonging to our English cottagers, where a few cabbages and carrots just peep up and grovel upon the earth between hedges, in square narrow beds, and where the tallest tree is a gooseberry bush: the vegetables of the negroes are all cultivated in their provision-grounds, which form their *kitchen*-gardens; but these are all for ornament or luxury, and are filled with a profusion of oranges, shaddocks, cocoa-nuts, and peppers of all descriptions. (55–56)

Lewis suggests that the slaves on his plantation live in an abundant paradise, a virtual garden of Eden. A distinctly different set of imperial dynamics are at work in this description, in which slavery is sentimentalized. Here the colonizers are having it mostly their own way as black West Indians stay in their place, providing pleasing and tasteful diversions to English eyes. But if left to their own devices, Trollope later cautions, they fail to produce such pleasing effects. Lewis' description also erases slave labour from the picture; we encounter no slaves in fields working under harsh and inhumane conditions in an account which favourably compares the situation of slaves to that of English peasants.[15]

In *The English in the West Indies*, Froude also figures the black West Indians as inhabiting a version of the biblical garden of Eden: "They live surrounded by most of the fruit which grew in Adam's paradise" (42). But he notes, with not a little displeasure, that "pioneering blacks were clearing patches of forest for their yams and coffee" (148). Similarly, Trollope complains:

> Port Antonio [in Jamaica] was once a goodly town, and the country round it, the parish of Portland, is as fertile as any in the island. But now there is hardly a sugar estate in the whole parish. It is given up to the growth of yams, cocoas, and plantains. It has become a provision-ground for negroes, and the palmy days of the town are of course gone. (25)

---

[14] Matthew Gregory Lewis, *Journal of a Residence Among the Negroes in the West Indies* (London: John Murray, 1845): 31.

[15] Such comparisons feature throughout pre- and post-emancipation travel narratives.

Not only does Britain have competition from the United States, it also must fight competition from its former slaves. "Nature and the dark race," Froude laments, "had been simply allowed to resume possession of the island"[16] (48). He warns his readers again and again that the West Indies could become another black republic like Haiti, "where they eat the babies" (50):

> On the hill-sides were the garden farms of the blacks, which were something to see and remember. They receive from the Government at an almost nominal quit rent an acre or two of uncleared forest. To this as the first step they set light; at twenty different spots we saw their fires blazing. To clear an acre they waste the timber on half a dozen or a dozen. They plant their yams and sweet potatoes among the ashes and grow crops there till the soil is exhausted. Then they move on to another, which they treat with the same recklessness, leaving the first to go back to scrub. (210)

Referring to the Afro-Caribbeans as "unthrifty," Froude implies, along with Trollope, that, unlike the British, they have failed to understand the nature of the West Indies and are thus unfit custodians of the islands (193). He mentions that the soil is exhausted and consequently rendered "useless and unprofitable" by their practice of burning to clear grounds; it is they, not the colonizers, who are the despoilers of the environment. Their crime, however, according to Froude's imperialist reasoning, is not against the environment, but against the edicts of mercantile capitalism. Black West Indians are not accused of environmental degradation or mismanagement, but of the mismanagement of resources and capital. Froude's home audience, whom the writer considers the rightful heirs to the West Indies, are informed that the descendants of England's slaves are growing rich and urges Britons to act fast, as "the smoke of their clearances showed where they were at work" (219). The West Indies, Froude declares, "were intended to be homes for the overflowing numbers of our own race, and the few that have gone there are being crowded out by blacks from Jamaica and the Antilles" (318).

Froude is particularly interested in ensuring that British imperialism in the Caribbean continues into the twentieth century. Notably, for all their attempts to locate alternative forms of industry for the West Indies, Froude and Trollope, despite at times bemoaning the absence of suitable hotel facilities,[17]

---

16 The island he is referring to here is Grenada.

17 Describing the Claredon, a hotel in British Guiana, Trollope writes: "It is a rickety, ruined, tumble-down, wooden house into which at first one absolutely dreads to enter, lest the steps should fail and let one through into unutterable abysses below" (132).

cannot as yet envisage tourism – that most effective and highly lucrative twentieth-century version of colonialism, keeping the "natives" in their place, and wreaking its own form of environmental havoc. In the twentieth century, the new imperialists from North America, who are the major shareholders in the Caribbean tourist industry, have replaced the older-style British imperialist[18] and transformed the face of the landscape in ways as dramatic as the earlier sugar industry.[19]

WORKS CITED

Amphlett, John. *Under a Tropical Sky: A Journal of First Impressions of the West Indies* (London: Sampson Low, Marston, Low, & Searle, 1873).

Andrews, Malcolm. *The Search for the Picturesque: Landscape Aesthetics and Tourism in Britain, 1760–1800* (Stanford CA: Stanford UP, 1989).

Bermingham, Ann. *Landscape and Ideology: The English Rustic Tradition, 1740–1860* (London: Thames & Hudson, 1986).

Brantlinger, Patrick. *Rule of Darkness: British Literature and Imperialism, 1830–1914* (Ithaca NY: Cornell UP, 1988).

Froude, James Anthony. *The English in the West Indies* (London: Longmans, 1888).

Kincaid, Jamaica. *A Small Place* (New York: Plume, 1988).

Lewis, Matthew Gregory. *Journal of a Residence Among the Negroes in the West Indies* (London: John Murray, 1845).

Lunenfeld, Marvin, ed. *1492: Discovery, Invasion, Encounter* (Sources in Modern History series; Lexington MA: D.C. Heath, 1991).

---

[18] In her dismantling of the tourist gaze in *A Small Place*, Jamaica Kincaid suggests that tourism in the Caribbean replicates in significant ways the system of relations under slavery. Slavery, the writer implies, provides the foundational features and terms upon which the relationship between the tourist and the landscape and people of Antigua are based. Kincaid suggests that, as in the days of slavery, vast areas of Antigua (her specific point of focus) and, by extension, the islands in the Caribbean are once again the property of "absentee landlords," but in the present-day West Indies tourism has replaced sugar production as the region's principal industry.

[19] John Amphlett, in his travelogue *Under a Tropical Sky*, contemplates the viability of tourism as an industry in the Caribbean and writes: "I see no reason why the West Indies should not become as favourite a health resort during our cold winter as the South of France. [...] There are hotels, but they are certainly not up to the requirements of the fastidious; but on the other hand it only requires more visitors to bring about the establishment of better hotels"; Amphlett, *Under a Tropical Sky: A Journal of First Impressions of the West Indies* (London: Sampson Low, Marston, Low, & Searle, 1873): 178.

Nugent, Maria. *Lady Nugent's Journal* (Kingston, Jamaica: Institute of Jamaica, 1966).

Pratt, Mary Louise. *Imperial Eyes: Travel Writing and Transculturation* (London: Routledge, 1992).

Tiffin, Chris. "Progress and Ambivalence in the Colonial Novel," in *Re-Siting Queen's English: Text and Tradition in Post-Colonial Literatures*, ed. Gillian Whitlock & Helen Tiffin (Cross/Cultures 7; Amsterdam & Atlanta GA: Rodopi, 1992): 1–9.

Trollope, Anthony. *The West Indies and the Spanish Main* (1859; Gloucester: Alan Sutton, 1985).

Polluted River or Goddess and Saviour?

The Ganga in the Discourses of Modernity and Hinduism

MEENAKSHI SHARMA

IN RAJA RAO'S AUTOBIOGRAPHICAL NOVEL *The Serpent and the Rope* (1960), Rama, the young Brahmin studying philosophy in France, says:

> The Ganges dissolves all sin. Even the ashes of the dead that the fire has burnt must dissolve in the Ganges and have absolution.
>
> *Sakala kalusha bhange svarga sopana sange*
> *Taralata range Devi Gange prasida*
> Dissolver of blemishes
> Companion of the Waters
> Dancing and sparkling Ganges I worship.
>
> Benares is everywhere where you are, says an old vedantic text, and all waters are the Ganges."[1]

For Rama, the Ganges is wisdom incarnate: "Whether young or older in years the Ganges is ever so knowing, so wise. If wisdom became water the Ganges would be that water, flowing down to the seven seas" (41). As they go up to Hardwar, Rama feels that even the train chants mantras: "*Namas-*

[1] Raja Rao, *The Serpent and the Rope* (1960; New Delhi: Orient, 1968): 382–83. Further page references are in the main text.

thethu Gange twadange bhujange" (40). His own feelings for the "grave and knowing river" are those of "pagan" worship" (23). For Europeans such as Catherine and his own wife Madeleine, the Ganges evokes images of floating human flesh, the pyres of the dead and crocodiles; but when Rama takes a boat ride on the river near Benares he sees none of these:

> Flowers floated downstream, and now and again we hit against a fish or a log of wood. Sometimes too a burnt piece of fuel from some funeral pyre would hit against the oars of the boat. People say there are crocodiles in the Ganges, and some add that bits of dead bodies, only half-burnt, are often washed down by the river. But I have never seen these myself. (23)

Rather than ascribing the horror of death to the Gangetic landscape, Rama believes that Benares and the Ganges attest to the fact that "death is as illusory as the mist of the morning. The Ganges is always there" and "the dead do not die nor the living live. The dead come down to play, on the banks of the Ganges, and the living [...] live in the illusion of a vast night and a bright city" (11, 22). When he acknowledges impurities in the river, it is not physical pollutants and corpses but the burden of worldly sorrows, sins and spiritual impurities: "She it was, from age to age, who had borne the sorrows of our sorrowful land. Like one of our own mothers, Ganga, Mother Ganga, has sat by the ghats, her bundle beside her. What impurity, Lord, have we made her bear" (33).

A Western account of the river such as Eric Newby's *Slowly Down the Ganges* (1966), despite sympathetic consideration of Hindu sentiments and beliefs about the river, cannot match that of Rao, who combines in his representation of the river both popular beliefs and rituals, and the high philosophy of *advaita* vedantism. For Rama, the Ganges is not just a body of water about which *other* people have certain beliefs, but a repository of stories, legends and myths internalized through a way of life in which there is no watertight divide between past and present, or between real life and legends, myths and religion. The Ganges is invested with myriad associations as a sacred goddess and mother – "the absolver of sins" – from the myth of Ganga's heavenly origins, Upanishadic stories of sages, incidents from Buddha's life, and of the poet-saints Tulsidas and Kabir, and the Brahmin-poet Jagannath Bhatta who married Shah Jehan's daughter, to the philosophy of Sankara wherein "all brides be Benares born" and "the Ganges flowed everywhere" (216, 218). For every valuable experience of life, the Ganges provides the fitting image. It is a metaphor for Rama's relationship with Savithri:

> Whatever I gave her she accepted, as the Ganges receives the waters of the Himalayas, that go on down to the sea and come again as the white flakes of snow, then blue again, the ice melts and once more the Ganges takes the waters down to the sea – so we gave love to each other. (170)

The significance of the Ganga down the ages is not something confined to history books but a lived experience connecting past to present – "Truth is the Himalaya, and Ganges humanity. That is why we throw the ashes of the dead to her. She delivers them to the sea, and the sun heats the waters so that becoming clouds, they return to the Himalayas" (35). The river and the mountains become kith and kin to Rama and his companions, who leave "with immeasurable pain, as though we had been visiting some venerable relations and had to leave them, with a broad kumkum on our faces and their hands on our heads" (112).

Eric Newby, however, despite recording the 108 names of the Ganges and its mythic origins, does not have access to the kind of osmotic faith in its holiness which Hindus absorb almost unconsciously. In 1963–64, Newby journeyed with his wife down the Ganges from Hardwar to the Sandheads just offshore from the Bay of Bengal. He describes the Ganges as a "great river," even though "in most standard works of reference the Ganges does not even rate an entry in the tables which list the great rivers of the world." "It is great," he continues,

> because, to millions of Hindus, it is the most sacred, most venerated river on earth. For them it is *Ganga Ma* – Mother Ganges. To bathe in it is to wash away guilt. To drink the water, having bathed in it, and to carry it away in bottles for those who have not had the good fortune to make the pilgrimage to it is meritorious. To be cremated on its banks, having died there, and to have one's ashes cast on its waters, is the wish of every Hindu. Even to ejaculate "*Ganga, Ganga*," at the distance of 100 leagues from the river may atone for the sins committed during three previous births.[2]

Newby appends the *Gangastottara-sata-namavali*," the 108 names of the Ganges, and even mentions several in the Introduction: "The Pure, The Eternal, The Light Amid the Darkness, The Cow Which Gives Much Milk, The

---

[2] Eric Newby, *Slowly Down the Ganges* (1966; London: Hodder & Stoughton, 1989): xvi. Further page references are in the main text.

Liberator, The Destroyer of Poverty and Sorrow, The Creator of Happiness" (xvi). Delving into history, he records that the Ganges was not always regarded so highly and Aryan invaders were more impressed by the Indus, only later according "*Ganga* the highest position, as *Sursari,* River of the Gods – perhaps because they had found out what European scientists discovered later: that its water had remarkable properties" (xvi). When Newby mentions these extraordinary qualities he does so without apparent scepticism, seeming to subscribe to the general beliefs of the Hindus:

> Bottled, it will keep for at least a year. At its confluence with the River Jumna which, particularly at the time of the great fair which takes place there every January, contains dangerous numbers of *coli*, the Ganges itself is said to be free of them. At Banaras thousands drink the water every day at bathing places which are close to the outfalls of appalling open drains. They appear to survive. The presence of large numbers of decomposing corpses seems to have no adverse effect on it. [...] Taken on board sailing ships in the Hooghly at Calcutta it is said to have outlasted all other waters. (xvi)

Newby and his wife try it themselves: "before setting off we were advised that when we wanted to make tea sedimentation could be accelerated by stirring it with a stick of alum; but we never did this" (xvi). He notes that the water of the Ganga had been described by everyone from Mr Nehru to the old sadhu at Hardwar as something from which no harm could come, adding his own favourable opinion of the river: "It is certainly responsible for more good than evil" (xvii). Hindu practices and customs are recorded sympathetically and, in an uncanny identification with the Hindu desire to die by the Ganga, he recounts how, along the way, sick and on land, his wife "could think of nothing else but the clean sandbanks and the solitude of the Ganges, betraying similar feelings to those of a Hindu dying after a long pilgrimage just out of sight of the river" (98).

Newby also cites Lord Dhanwantri, who opined that "When the body is afflicted by senility and diseases, the holy water of Mother Ganga is the medicine, and Lord Narayana, from whose feet Ganga emanates, is the great physician" – the traditional Hindu incantation to holy rivers during a bath, and Swami Sivananda's *Mother Ganges* (76). Yet, despite its remarkably sympathetic tone, Newby's account remains that of an outsider – a Western onlooker and commentator – who assumes, perhaps unconsciously, the role of a colonial traveller-anthropologist:

> This was no uncharted river. Millions lived on its banks, regarding it as an essential adjunct without which their existence would be unthinkable, if not impossible; bathing in it; drinking it; washing their clothes in it; pouring it on to their fields; dying in it; being taken into its bosom by it and being borne away. (4–5)

Moreover, his modern Western sensibility makes him record unequivocal admiration for Captain Proby Cautley whose brain-child – the Ganges Canal – brought prosperity to the Doab between the Ganges and the Jumna which had suffered from frequent famines. The other side of development – the depletion of the Ganga and its reduced capacity for self-purification – is not described by Newby. Similarly, there is no mention of the effect of deforestation and industrialization – especially in the context of colonialism – on the condition of the river.

Durgacharan Ray's Bengali text *Debganer marttye agaman* – cited in Partha Chatterjee's introduction to *Texts of Power* (1995) – brings us to a crucial element in the discourse of the Ganga: colonialism. For the Indian inhabitants, the gods arrived in Calcutta in 1880 to be met by Ganga, who "complained bitterly to them about the indignities she was being forced to suffer at the hands of the English, the new rulers of the country."[3] Brahma heard the woeful tale of his daughter and saw, with distress, the massive interventions in the world he had created:

> The river was so thick with boats, large and small, of all descriptions, that the water was hardly visible. The air was heavy with the sounds of horns, sirens, and whistles emanating from vessels. There were English sailors perched on the masts of the large ships. On the opposite bank of the river, bathers had lined up like rows of ants on the ghats (riverbank steps). Beyond them, in a line stretching from one end of the horizon to the other, stood the great mansions of the city, interspersed at points by huge chimneys belching smoke into the sky. (1)

A weeping Ganga catalogued her suffering to her father:

> The new rulers of the country are not content with having me bound in chains. They force me like a slave to carry their huge ships. If I say it's too hard for me, they tug the boats all over me. There are carriages and men crossing that bridge at all times of day and night. I don't have a minute's rest. And now they've put up all kinds of factories on my banks: the noise and the smoke are unbearable. I've

[3] Partha Chatterjee, Introduction, "The Disciplines in Colonial Bengal," to *Texts of Power: Emerging Disciplines in Colonial Bengal*, ed. Chatterjee (Minneapolis: U of Minnesota P, 1995): 1. Further page references are in the main text.

> never seen rulers like these in my life. When they need land, they grab it from me. Just look how far they have filled up the land near the mint. They tax the boats, they tax the fish, they tax the corpses, they even tax my waters. (2)

The modernization and development wrought by the British impressed even the gods, who could not perceive that the "wonderful inventiveness" of the colonial rulers was "not unrelated to the humiliations they were inflicting on [them] as well as the people of the country" (3). In this representation, it is Western modernity and its adjuncts – industrialization and material progress – which are responsible for the denigration of indigenous cultural and religious traditions. The myth of cultural and intellectual superiority is consolidated by the strength of military / political superiority.

The river Ganga (which may metonymically be regarded as Hindu culture and tradition) is thus the site of a struggle between a colonialist modernity and an ancient and living tradition. The roots of this struggle lie in colonial rule, which superimposed Western paradigms of modernity upon the ancient and deep-rooted traditions of India. Colonial intervention for commercial gain disregarded – but could not entirely displace – the value of traditional cultural and religious beliefs (which, in the Hindu context, are inextricably interwoven). Particularly resilient were views about nature in a people for whom their way of life in many ways remained the same as that of their ancestors' *Aranya* [forest] society. One of the far-reaching legacies of colonialism is the exploitation of natural resources by intervention in precolonial patterns of resource use. At the same time, India's colonial past also left legacies in education, social dynamics, economics and polity resulting in an almost unavoidable co-option to Western ideologies of modernity, progress and development. This confluence necessarily resulted in a continuation of the processes set in motion in colonial times – processes further accelerated by globalization and liberalization during the closing decades of the twentieth century. However, the uneven and partial modernization that resulted from the dual heritage of ancient culture and colonial past allows postcolonial Indians to accept the inevitability – even the desirability – of forms of 'progress' in terms of industrialization, urbanization, optimalization of resources, and economic and social development while a "domestic" core of spiritual values and beliefs remains largely untouched.

But the historical conditions under which a Western modernity was imperfectly grafted onto the old and rich traditions of India resulted in an uneven, patchwork modernization, not only across India and its population, but also within individuals. It is because of this continuing coexistence of what are assumed (in Western thinking) to be mutually exclusive modes, that the hold of the Ganga's mythical and spiritual image has not been relinquished despite the 'modernization' of mind, institutions and land practices. In fact this very co-option of Indian tradition to Western paradigms of modernity has resulted, paradoxically, in an even stronger retention of spiritual traditions and values. The diversity and heterogeneity of Indian society has also contributed to the limited impact of what we might call a 'core' colonialism. As Ashis Nandy argues, colonialism had a deep impact on India and

> the economic exploitation, psychological uprooting and cultural disruption it caused were tremendous. But India was a country of hundreds of millions living in a large land mass. In spite of the presence of a paramount power which acted as the central authority, the country was culturally fragmented and politically heterogeneous. It could, thus, partly confine the cultural impact of imperialism to its urban centres, to its Westernized elite.[4]

Partha Chatterjee, in his analysis of the impact of colonialism on nineteenth-century intellectuals, distinguishes between the material and spiritual domains, and argues that the greater the compromises which have, perforce, to be made in the former, the greater the tenacity with which the sanctity of the latter is preserved and traditional modes of thought and conduct defended against corruption. While the West was accepted as having proved its superiority in technology and "its accomplishments [were] carefully studied and replicated," the latter were taken to bear the "essential" marks of cultural identity. "The greater one's success in imitating Western skills in the material domain, therefore, the greater the need to preserve the distinctness of one's spiritual culture."[5] A comparable model may be seen in the case of Japanese society, wherein a similar disjunction makes the modernity discernible in the practical / technical sphere paradoxically lead to greater conservatism and tradition in the cultural / spiritual sphere. In such societies, modernity and traditionalism seem to avoid confrontation by side-stepping each other.

While environmental historians have chronicled the exploitation of land and resistances to it – chiefly in the case of forests and dams – not much attention has been paid to how faith and traditional beliefs about sacred places have been brought into accord, however uneasily, with modernization.

---

[4] Ashis Nandy, *The Intimate Enemy: Loss and Recovery of Self Under Colonialism* (Delhi: Oxford UP, 1983): 31–32.

[5] Partha Chatterjee, *The Nation and Its Fragments: Colonial and Postcolonial Histories* (Princeton NJ: Princeton UP, 1993): 6.

Arnold and Guha, in their Introduction to *Nature, Culture, Imperialism* (1995), note that work on the great river systems of the Indian land mass would be very rewarding, as these "have exercised such a definitive influence on the natural environment, as well as the economic, social and religious life of the region."[6] A study of the impact of modernization and development on the Ganga, both in terms of its physical condition and on people's faith and belief about it, is thus extremely important for any understanding of the ways in which place – in this case a sacred body of water – reflects the particular relationship between land and people in Hindu culture.

The ecological history of the Indian subcontinent in terms of the exploitation and degradation of forest resources for economic gain is of crucial importance in understanding the larger context of the crisis of the Ganga river. But the case of this holy river also strikingly demonstrates the ways in which traditional beliefs about the divinity and unpollutability of the Ganga have allowed vested interests to wreak havoc on the river's ecosystem. Much of the media coverage of the pollution of the Ganga implies that the beliefs of the people contribute to the pollution of the river, and many observers express wonder at how such strong beliefs in the river's purity can coexist with an attitude of indifference to its physical condition. Increasingly, however, there is an awareness that reverence for the Ganga should not simply be dismissed but instead co-opted to support its regeneration. The failure of ambitious government projects to clean up the river suggests that its protection can be ensured only through the co-operation of local people, whose lives and faith are intrinsically interwoven with the river's fate.

Hindus believe that Ganga, the celestial river goddess, daughter of Brahma, came down to earth as a boon granted to Bhagirath in order to purify the souls of his 60,000 ancestors reduced to ashes by the wrath of an ascetic. The massive fall was broken by Shiva himself, who contained the force of the mighty river in his matted locks and allowed it to flow benevolently over the earth. For millions of Hindus the world over, belief in the divinity of the Ganga and her power to bring salvation is far stronger than any verifiable scientific law – bathing in its waters and immersing the ashes of their dead in it still have the aura of purity they have had for centuries. Western observers find it incredible that the heavily polluted waters of the river continue to be regarded in this way even by highly-educated Indians. Loke-

---

[6] David Arnold & Ramchandra Guha, ed. *Nature, Culture, Imperialism: Essays on the Environmental History of South Asia* (New Delhi: Oxford UP, 1995): 11.

nath Bandopadyaya's "The Goddess Who Came Down to Earth" acknowledges the impossibility of writing about the Ganga "since much of the river exists more in the imagination of the people [...] than in reality" (np). Although he laments the pollution of the river in modern times, he poses the question "what is more real in life, the reality, or imagination?" and answers it with "in the case of the Ganga at least, the answer to that question is: imagination" (np). Even though the water – a physical object – is used for worship, for purification rituals such as bathing, and for immersion of the ashes of the dead, the waters of the river retain their essence, untouched by its physical condition.

The belief of Hindus in the special qualities of the waters of the Ganga has, since colonial times, been repeatedly tested scientifically. Citing a study referred to by Swami Sivananda in *Mother Ganges*, Newby notes that "an unnamed Canadian professor, said to be of McGill University," had tested and confirmed its qualities.[7] A certain scepticism is obvious, although Newby is careful to make no explicit comments on the issue:

> A peculiar fact which has never been satisfactorily explained is the quick death, in three or five hours, of the cholera vibrio in the waters of the Ganges. When one remembers sewage by numerous corpses of natives, often cholera casualties, and by the bathing of thousands of natives, it seems remarkable that the belief of the Hindus, that the water of this river is pure and cannot be defiled and that they can safely drink it and bathe in it, should be confirmed by means of modern bacteriological research.[8]

In an article in *Geographical Magazine* entitled "Ganges Has Magical Cleaning Properties," Rakesh Kalshian cites the work of D.S. Bhargava, a scientist at the Roorkee Engineering University, who argues that the river is able to reduce b.o.d. levels much faster than other rivers, because there are unique microbes in it which help to clump suspended solids so that their increased mass allows them to sink, making the river particularly efficient at reaeration. The article also cites nineteenth-century studies by the British physician E. Hanbury Hankin, who found, on testing water from underneath the bodies of cholera victims thrown into the river, that cholera microbes died within three hours in Ganga water but continued to thrive in distilled water even after 48 hours. Another British physician, C.E. Nelson, reported that Ganga water taken from the Hoogly on ships returning to England remained fresh

---

[7] Newby, *Slowly Down the Ganges*, 193.

[8] *Slowly Down the Ganges*, 193.

throughout the voyage. French scientist Monsieur Herelle also reported that Ganga water just under victims of dysentery and cholera was free of germs.[9] Payal Sampat, in an article in *World Watch*, also refers to D.S. Bhargava's studies which reported that the Ganga decomposes organic waste fifteen to twenty-five times faster than other rivers, and that his findings "tally closely with those of the government's Central Board for the prevention and Control of Water Pollution in New Delhi. This finding has never been fully explained."[10]

There is a note of puzzlement in such reports. The explanations offered for the amazing findings range from the presence of unnamed micro-organisms that help to decompose pollutants, to qualities shared by rivers of this size generally. There is consensus, however, that perhaps due to its *particular* flora and fauna, depth and flow, its many tributaries, as well as the role played by the periodic monsoons in flushing out impurities, the river has been extraordinarily efficient in self-purification. Originally the site of small-scale organic farming, without large cities or large-scale industries to cause massive pollution, the river coped with human-created effluents with remarkable efficiency, thus justifying the faith in its holy waters. However, under the combined onslaught of severe ecological stresses, those life-supporting properties which justified the reverence of Indians down the ages are in danger of collapsing. Such stresses include the depletion of its volume by canals; urbanization and its demands for drinking water; sewage disposal; industrialization with blatant disregard for environmental responsibilities; and erosion, silting and chemical fertiliser run-off. As Sampat writes, "for a long time, the river seemed impervious to damage; its enormous volume of water diluted or decomposed waste very rapidly, and the annual monsoons regularly flushed it out. [...] But with 20th century pressures of burgeoning population and industrial growth, the Ganges is teetering under the burden placed on its cleansing capacities."[11]

The devotee-users who are represented as the villains – those who allegedly do not allow the river to be cleaned by denying the possibility of its pollutability – are more precisely the victims of modernization and development. Their disregard for the river's polluted state is not so much blindness to

---

[9] Rakesh Kalshian, "Ganges Has Magical Cleaning Properties," *Geographical Magazine* 66.4 (1994).

[10] Payal Sampat, "The River Ganges' Long Decline," *World Watch* 9.4 (1996).

[11] Sampat, "The River Ganges' Long Decline."

its physical condition as a tribute to the resilience of a faith which, while encouraging reverence, does also make for small, personal adjustments in the use of its waters. Newby records that devout Hindus are careful about the exact point in the river's course up to which they consider its waters to be "pure."[12] A recent newspaper report mentions that Veer Bhadra Mishra begins his day by taking a holy dip in the Ganga, "though these days, he holds his nose and doesn't swallow any water."[13] However, traditional attitudes to the river, which are a unique combination of the metaphorical and the actual, are partly responsible for creating the situation whereby the physical desecration of the river's waters can sometimes be overlooked. While ordinary people may find themselves helpless in combating the degradation of the river, it cannot be assumed that they oppose its cleansing. Historically, the degradation of the river's ecosystems has been due not to pilgrims but to macro-level projects such as massive deforestation, construction of large dams and canals, and fundamental changes such as industrialization and urbanization – all of which ordinary people could neither halt nor combat. In the postcolonial context, the modernization of industry and of the economy, the pragmatics of 'development', and the imbrication of the overall Indian consciousness in Western modernity has resulted in a scenario whereby the traditional regard for the Ganga has not weakened, but it has to some degree become immune to the effects – including deleterious ones – of modernization on the holy river.

Fran Peavy, an American social-change counsellor working for Veer Bhadra Mishra's *Sankat Mochan* Foundation's project to clean up the Ganga at Varanasi, writes in an article in *Whole Earth Review* (1995) about how she set out to find out how Indians explained the pollution to themselves. She does not report finding resistance to cleaning up the river nor denial that it was polluted – "over and over I heard something like, 'the river is holy, but she is not pure. We are not taking care of her the way she needs us to'" and "people often said, 'I see the problem but others don't'."[14] Although she writes that it was not wise to say up-front to Indians that the "river is polluted" – which would be equivalent to saying to a Westerner that "your mother is a

[12] Newby, *Slowly Down the Ganges*, xvi.

[13] Namita Devidayal, "Öm Clean Gangaya Namah," *The Times of India* (26 February 1999).

[14] Fran Peavy, "Questions for the Ganges: Culturally Sensitive Strategies for Cleaning Up a Revered Indian River," *Whole Earth Review* 86 (1995).

whore" – she found that people were ready to employ the strategies of change embedded in Indian culture such as "satyagraha, fasting, direct action, pressuring civic leaders, citizens' assemblies, marches to the capital [...] to clean up their holy river" and enthusiastic about devising strategies for spreading awareness about the river's condition.[15]

Studies on the Ganga almost invariably mention that it is one of the most polluted rivers in the world, enumerating the sources and nature of the pollutants. Often ignored is the fact that the roots of the problem lie in colonial times, when economic gains were achieved at the cost of the ecological balance maintained by indigenous peoples through centuries of harmonious coexistence with nature. As Richard Grove writes, it is only now being realized that

> in many ways the business of empire, for most of the colonised, had far more to do with the impact of different modes of colonial resource control and colonial environmental concepts, than it had to do with the direct impact of military or political structures.[16]

Grove delineates the process of ecological transition which "largely [...] followed upon the spread of a European capitalist system over the globe with the corresponding penetration of a Western economic process beyond as well as within the colonial context" as

> transformation in the nature of the tenurial relationships between people, forests and other non-arable land [... which] involved, in essence, a transition away from locally evolved man-land relations towards direct private property status or to direct state control. These changes have often involved a growing exploitation of the landscape for commodity production and a corresponding erosion in customary controls and common property rights or conventions.[17]

Although it does not do to romanticise the past as one of perfect balance between humans and the land, it is important to remember that in the Indian context a well-entrenched culture of living harmoniously with nature, and a religious mythology that peopled the forests and rivers with gods and goddesses, allowed for a more sustainable use of natural resources for meeting human needs. "To the ancient peoples of Mesopotamia or the Ganges valley,

---

[15] Peavy, "Questions for the Ganges."

[16] Richard H. Grove, *Ecology, Climate and Empire: The Indian Legacy in Global Environmental History, 1400–1940* (New Delhi: Oxford UP, 1998): 3.

[17] Grove, *Ecology, Climate and Empire*, 179.

there was possibly no greater crime than the desecration of a river."[18] As Bittu Saghal writes in the *Sunday Observer*,

> 'the Forest is the Mother of the River'. [...] By worshipping and protecting forested water sources we managed through the ages not only to harvest water, but also to temper the sheer force of the monsoon. But we have forgotten such lessons in our haste to ape the industrial North.[19]

J. Bandopadhyay and Vandana Shiva describe tree worship as "a cultural response of a civilisation which perceived the ecological value of the conservation of forests":

> The sacredness of trees was the cultural category that ensured the material conservation of essential life support systems. [...] The metaphor of the sacred tree and the sacred forest has been a powerful metaphor in mobilising entire societies for sustainable use of forest resources and economic development. Sacredness under these conditions becomes a useful materialist category, not a spiritual one, since it ensures maintenance of the material basis of life."[20]

Their aphorism – "trees as symbols of cash and profit are destroyed. Trees as sacred are protected and used prudently"[21] – is proven by the experience of colonial rule, wherein the profit motive led to the displacement of "the perception of forest ecosystems as having multiple functions for satisfying diverse and vital human needs" and the establishment of "one-dimensional scientific forestry [...] which had as its only objective the maximisation of the production of commercially valuable timber and wood while ignoring other ecological and economic objectives for the utilisation of forest resources."[22]

The post-independence path of India's development has followed along the lines set down by its colonial masters. As Gadgil and Guha argue, "British colonial rule marks a watershed in the ecological history of India. The country's encounter with a technologically advanced and dynamic culture gave rise to profound dislocations at various levels of Indian society." Not only did their political victory equip the British "for an unprecedented intervention in

[18] Sampat, "The River Ganges' Long Decline."

[19] Bittu Sahgal, "Water, Water Everywhere...," *Sunday Observer* (22 March 1998).

[20] J. Bandopadhyay & Vandana Shiva, Introduction to *Forests in India's Heritage*, ed. Bandopadhyay & Shiva (Dehra Dun: Research Foundation for Science, Technology and Natural Resource Policy for Chipko Movement, nd): 1.

[21] Bandopadhyay & Vandana Shiva, Introduction, 1.

[22] Bandopadhyay & Vandana Shiva, Introduction, 20.

the ecological and social fabric of Indian society, but, by exposing their subjects to the seductions of the industrial economy and consumer society, the British ensured that the process of ecological change they initiated would continue, and indeed intensify, after they left India's shores."[23] In fact,

> the tasks [British imperialism] had left unfinished were enthusiastically taken up by the incoming nationalist elites, whose unswerving commitment to a resource-intensive pattern of industrialization has only intensified the processes of ecological and social disturbance initiated by the British.[24]

Yet, while the impact of westernization on polity, economy and infrastructure could not be resisted, in the cultural domain the larger portion of Indians were not culturally dislocated from their traditions and beliefs. In contemporary times this is evident in the kinds of attitudes to the Ganga – Indians who live and enjoy / suffer the fruits of modernization and are, in various ways, inextricably imbricated in it – nevertheless display the vitality of traditionalism and faith.

Jawaharlal Nehru, independent India's first Prime Minister and the architect of its development policies in the first two decades after independence, was of the view that "we cannot stop the river of change or cut ourselves adrift from it and, psychologically, we who have eaten the apple of Eden cannot forget the taste and go back to primitiveness."[25] Nehru thus turned his back upon the Gandhian philosophy of strengthening the village economy and deploying traditional, precolonial modes of development which posited the village community as the basic unit. As Kamla Chowdhury notes, at Independence Nehru made the choice of "a strong industrial (and military) state" and "Gandhi's concerns of village development, village industries, economic and political decentralisation, the empowering of local people in relation to local resources were set aside."[26] Paradoxically, although the success of the Indian Congress in enlisting a very broad-based support in the peasantry was based on Gandhi's appeal to precisely this group, it came to be increasingly influenced by the capitalist class rather than by Gandhi, who rejected

---

[23] Madhav Gadgil & Ramachandra Guha, *This Fissured Land: An Ecological History of India* (Delhi: Oxford UP, 1993): 5, 118.

[24] Gadgil & Ramachandra Guha, *This Fissured Land*, 242.

[25] Quoted in Kamla Chowdhury, *Industrialisation, Survival and Environment: A Dialogue on Development* (INTACH Environment Series 8; New Delhi: INTACH, 1989): 2.

[26] Chowdhury, *Industrialisation, Survival and Environment*, 3.

"modern industry in its totality…[and] pleaded for a revival of the organic village communities of the pre-colonial and pre-industrial past."[27]

It is therefore fitting that Gandhi has emerged as "the patron saint of the environmental movement" in India;

> leading activists have invoked his ethic of self-restraint, his attacks on consumerism, and his celebration of village society as providing building blocks for the construction of an environmentally and socially harmonious alternative to modern industrial development.[28]

> In the path of economic development eventually charted by the Indian nation, the Mahatma's ideals were made redundant with a quite alarming rapidity. Most Indian nationalists drew a wholly different conclusion from the colonial experience, arguing that India's subjugation was a consequence of its intellectual and economic backwardness. In [sic] this perspective, as contrasted with the dynamic and progressive West, India was a once-great civilization that had stagnated under the dead weight of tradition. Its revitalization could come about through an emulation of the West.[29]

The standard assumptions of modernization theories underpinned the conviction that "through rapid industrialization and urbanization India could 'catch up' with the West" and, not surprisingly, this vision grew in strength with the rise of the capitalist class. It was, therefore, the second alternative that was followed and the vision of "industrialize or perish" was institutionalized in the Second Five Year Plan – "underlaying a strategy of imitative industrialization was the adoption of most 'modern' technologies, with little regard for their social or ecological consequences."[30] While the filtration theory of education and acculturation employed by the colonial rulers had produced uneven results, this trickle-down approach to development in independent India has reinforced and created effects that are even more uneven. It is not surprising, therefore, that India offers the unique spectacle of a society where many centuries apparently co-exist in the most puzzling configuration. With the liberalization into which India, like other developing nations, has been lured by International Monetary Fund and World Bank policies, the process of modernization has now been even more firmly consolidated.

---

27 Gadgil & Guha, *This Fissured Land*, 181.

28 Arnold & Ramchandra Guha, ed. *Nature, Culture, Imperialism*, 18.

29 Gadgil & Guha, *This Fissured Land*, 183.

30 *This Fissured Land*, 184.

The most important causes of the river's degraded ecosystems have been deforestation, which does immense damage in terms of causing erosion/siltation and affecting monsoons; barrages and large dams; canals, which cut down on volume and flow of water, thus limiting the capacity to dilute wastes; and pollution, chiefly comprising domestic waste, chemical fertiliser and pesticide run-off and industrial discharges. While industrial wastes are smaller in volume, constituting about fifteen percent of the total, they have a far more insidious impact on the river.[31] In *The State of India's Environment 1982: A Citizens Report*, it is reported that in areas largely free of industrialization where the river has had to deal with sewage discharges alone, the water was relatively clean. It is thus ironic that the Ganga Action Plan focused on sewage waste rather than industrial wastes – a politically cautious approach which did not offend powerful interests. On the basis of various studies, the 1982 *Report* reconfirmed the great self-purifying capacity of the Ganga, but cautioned that this was not cause for complacency since "the natural defenses have been breaking down and urgent steps to halt the degradation are called for just upstream of the towns and cities."[32] It regretted that despite "every study since the early 1960s" having underlined the urgent need for controlling the Ganga's pollution, little has been done to implement the recommendations, with the result that

> in the public mind the Ganga flows on as of yore cleansing the sins of generations. Environmental awareness remains confined to committees and groups of concerned scientists. Shiva is no longer required to protect the Earth from the Ganga; now the Ganga itself requires protection from the violence of the people.[33]

The "violence of the people" is the profit-driven exploitation of natural resources. Veer Bhadra Mishra rightly distinguishes between need and greed: "we may exploit the Ganga for our needs, but not for our greed."[34] This position epitomizes the distinction between, on the one hand, the mythological origins of the river in answer to man's need for salvation and, on the other, the kind of exploitation / abuse initiated during colonial times. While the

---

[31] Sampat, "The River Ganges' Long Decline."

[32] Centre for Science and Environment, *State of India's Environment 1982: A Citizens'Report* (New Delhi: Centre for Science and Environment, 1982): 25–26.

[33] Centre for Science and Environment, *State of India's Environment 1982*, 26.

[34] Quoted in Ramesh Menon, "Ganga Clean-Up: Monumental Failure," *India Today* (15 January 1997): 120.

former offers a living relationship with the river, the latter is an exploitative practice based on greed.

To combat the pollution of the Ganga, then Prime Minister Rajiv Gandhi launched the Ganga Action Plan (GAP) – an ambitious programme for cleaning up the Ganga – in 1985. The apparent clash of traditionalism and modernity is reflected in the persistent representations of Rajiv Gandhi as "modern" and "agnostic," while traditional Hindus are depicted as opponents of the clean up. Mary Anne Weaver, writing in *Forbes*, describes Rajiv Gandhi as a scientific-tempered "agnostic" who is "ready for the next century," while "many of his countrymen" are not. Although she mentions that the "potentially fatal opposition" to Rajiv's programme of cleaning up the Ganges comes from vested interest groups such as political and religious leaders as well as local industrialists, she goes on to imply that it is the traditional and religious-minded Hindus who are resisting the schemes of the modern/agnostic Prime Minister. Although she quotes Veer Bhadra Mishra as saying, "we can't tell people the river is impure. It offends their religious sensibilities," she does not go on to explain that Mishra – a professor of hydraulic engineering and Hindu high priest – has been at the forefront of the movement in Varanasi to mobilise the people in the cause of saving the river and for devising participatory methods in conjunction with locally suitable techniques of waste-treatment. Weaver's implication is that despite being a high priest, Mishra's technical qualifications distinguish him from "the people" who unquestioningly accept the river's unpollutability and regard admitting its polluted state as sacrilegious.[35]

However, it is not the "people" who have resisted the clean-up on the grounds of unpollutability but vested interest groups who have cynically employed the rhetoric of tradition and faith to maintain the *status quo*, allowing for maximum profit without responsibility for ecological degradation. The people's support for a cleaner Ganga is obvious from the vociferous dissatisfaction with the performance of the Ganga Action Plan, and the heartening results of participatory schemes. After all, those who continue to revere the Ganga and those whose livelihood binds them closely to it only stand to benefit from any improvements. As Anuradha Raman writes in a report in *The Pioneer*, "people living by the side of the Ganges are perhaps as en-

---

[35] Mary Anne Weaver, "Roiling the Holy Waters," *Forbes* (5 May 1986).

dangered as any other species facing a threat to its existence."[36] In the Gangetic region, farmers complain of deteriorating productivity of land, while water-related ailments such as amoebic dysentery, gastro-enteritis, tapeworm infestations, typhoid, cholera and viral hepatitis are extremely common. In December 1997, while preparations were being made for the coming Mahakumbha Mela which brings millions of pilgrims to the banks of the Ganga, scientists of the National Environmental Engineering Research Institute made the alarming discovery of very significant contamination of the Ganga at Hardwar with antibiotic-resistant bacterial and viral pathogens. Industrial effluents were suspected to be responsible for this situation.[37]

Although the GAP was almost unanimously declared a failure, with even the advisor to the GAP directorate admitting technical flaws and poor implementation by state agencies, the reason for the failure is not people's resistance. It is rather due to lack of political will both in implementation of guidelines and financing of schemes, as well as in not tackling the contentious problem of industrial effluents far more ecologically damaging than domestic waste disposal. As media reports indicate, there has been considerable resentment from people affected by the botched clean-up schemes, and concern of NGOs and ecological groups over the failure of high-cost schemes under GAP leading to a High Court directive in 1998 for a high-level audit of the programme. The Government's efforts to clean up the river have been under fire for tardy and faulty implementation, waste of money, bureaucratic inefficiency, diversion of funds and large-scale corruption, political apathy and capitulation to pressure from vested-interest groups. There has also been considerable resentment at the non-involvement of local people in the planning and implementation of GAP. A report in the *Hindu* notes that the High Court directed

> that the GAP should be implemented by involving local bodies which should be authorised to construct, maintain and operate the treatment plants, and undertake the river construction work at their own level with the help of the Government.

The report adds that "since people's participation has become the theme song for successful execution development or anti-pollution projects, the MoEF

---

[36] Anuradha Raman, "Despite GAP-I, Polluted Flows the Ganga," *The Pioneer* (7 January 1999).

[37] Bharat Dogra, "Pollution of a Holy River," *TWN Features* (21 July 1998).

[Ministry of Environment and Forestry] has also finalised guidelines to enable public participation in river action plans."[38]

Such rethinking of people's role in the cleaning of the Ganga, and mobilization of affected people which has been achieved by movements such as the Sankat Mochan Foundation's Swatcha Ganga Campaign in Varanasi and the Ganga Mukti Andolan in Bihar, contradict those who argue that "the people" are not ready to accept that the holy river is polluted. A report in the *Hindu* quotes Rajiv Gandhi, who introduced GAP as "a plan not for PWD, but for the people of India," and recommends a return "to the commendable objectives of the GAP" by devising programmes to encourage public participation in the protection and cleaning of the river and its tributaries.[39] As Bharat Dogra notes, the crucial step is to link reverence to effective action: "How the faith of the people can be extended beyond empty (sometimes even harmful) rituals and linked to the protection and cleaning of the river is a question that needs to be examined seriously."[40] "A broad consensus should be created by involving community leaders around the belief that the real 'punya' or good deed exists in contributing to the cleaning and proper maintenance of rivers and other water sources, and not in outdated rituals which can sometimes even harm the rivers."[41]

While the myth of people's opposition to cleaning up the Ganga needs to be investigated, and the 'user pays' principle replaced by that of the 'polluter pays', the mythical image of the Ganga as mother and goddess can in fact be co-opted to a modern cause. The spiritual legacy of Hinduism is a vital force, one that has negotiated many social and political movements and exigencies to shape contemporary Indian sensibilities. This needs to be directed towards taking responsibility for the physical condition of the Ganga and the ecosystem, and towards forcing political action to arrest/reverse present ecological damage, whose roots, like much else in contemporary India, lie in the colonial past.

---

38 "Cleaning the Ganga," *The Hindu* (Madras; 2 June 1995).

39 "Cleaning the Ganga."

40 Bharat Dogra, *Protecting Ganga River: Linking Reverence to Effective Action* (Dehra Dun: Rural Litigation and Entitlement Kendra, 1998): 16–17.

41 Dogra, *Protecting Ganga River*, 21–22.

WORKS CITED

Anon. "Cleaning the Ganga," *The Hindu* (Madras) (2 June 1995).

Arnold, David, & Ramchandra Guha, ed. *Nature, Culture, Imperialism: Essays on the Environmental History of South Asia* (New Delhi: Oxford UP, 1995).

Bandopadhyay, J., & Vandana Shiva. *Forests in India's Heritage* (Dehra Dun: Research Foundation for Science, Technology and Natural Resource Policy for Chipko Movement, nd).

Centre for Science and Envoronment. *State of India's Environment 1982: A Citizens'Report* (New Delhi: Centre for Science and Envoronment, 1982).

Chatterjee, Partha. *The Nation and Its Fragments: Colonial and Postcolonial Histories* (Princeton NJ: Princeton UP, 1993).

——, ed. *Texts of Power: Emerging Disciplines in Colonial Bengal* (Minneapolis: U of Minnesota P, 1995).

Chowdhury, Kamla. *Industrialisation, Survival and Environment: A Dialogue on Development* (INTACH Environment Series 8; New Delhi: INTACH, 1989).

Devidayal, Namita. "Öm Clean Gangaya Namah," *The Times of India* (26 February 1999).

Dogra, Bharat. "Pollution of a Holy River," *TWN Features* (21 July 1998).

——. *Protecting Ganga River: Linking Reverence to Effective Action* (Dehra Dun: Rural Litigation and Entitlement Kendra, 1998).

Gadgil, Madhav, & Ramachandra Guha. *This Fissured Land: An Ecological History of India* (Delhi: Oxford UP, 1993).

Grove, Richard H. *Ecology, Climate and Empire: The Indian Legacy in Global Environmental History, 1400–1940* (New Delhi: Oxford UP, 1998).

Kalshian, Rakesh. "Ganges Has Magical Cleaning Properties," *Geographical Magazine* 66.4 (1994).

Menon, Ramesh. "Ganga Clean-Up: Monumental Failure," *India Today* (15 January 1997): 120.

Nandy, Ashis. *The Intimate Enemy: Loss and Recovery of Self Under Colonialism* (Delhi: Oxford UP, 1983).

Newby, Eric. *Slowly Down the Ganges* (1966; London: Hodder & Stoughton, 1989).

Peavy, Fran. "Questions for the Ganges: Culturally Sensitive Strategies for Cleaning Up a Revered Indian River," *Whole Earth Review* 86 (1995).

Raman, Anuradha. "Despite GAP-I, Polluted Flows the Ganga," *The Pioneer* (7 January 1999).

Rao, Raja. *The Serpent and the Rope* (1960; New Delhi: Orient, 1968).

Sahgal, Bittu. "Water, Water Everywhere...," *Sunday Observer* (22 March 1998).

Sampat, Payal. "The River Ganges' Long Decline," *World Watch* 9.4 (1996).

Weaver, Mary Anne. "Roiling the Holy Waters," *Forbes* (5 May 1986).

Ecotourism
A Colonial Legacy?

HELEN GILBERT

THE ARGUMENT THAT MODERN TOURISM frequently functions as a form of neocolonial enterprise is by now commonplace. John Frow, among others, has explored the ways in which tourism, at its most general level, sells a commodified relation to an ontological Other – be it a natural environment, a species of wildlife, or a foreign culture. This relationship, often manifest in ritualized practices such as sightseeing and souvenir-collecting, is secured via the aestheticization of various physical and cultural features of a tourist destination and by the commercialization of immaterial resources such as hospitality. The tourist's position as consumer assumes *a priori* access to sufficient capital to purchase an encounter with Otherness; hence, it follows that most tourists come from relatively affluent societies while it is the Others of Western modernity who are often called upon to supply the requisite quotient of exotica for the collective tourist gaze. As Frow maintains, the logic of tourism thus becomes "that of a relentless extension of commodity relations and the consequent inequalities of power between centre and periphery, First and Third Worlds, developed and underdeveloped regions, metropolis and countryside."[1]

As a relatively new form of leisure activity – at least under its current nomenclature – ecotourism has sought to define itself in opposition to the

[1] John Frow, "Tourism and the Semiotics of Nostalgia," *October* 57 (1990): 151.

kind of mass tourism that Frow's analysis implicitly decries. In its purer forms, ecotourism is even premissed on behaviours and subject–object relations which are designed to break the relentless cycle of inequality that commodification perpetuates. The recent rapid growth in this form of travel, especially in developing countries and those regions of developed nations populated largely by indigenous minorities,[2] suggests the rhetorical force of ecotourism as a discursive field and its appeal to both ethical-minded tourists and potential host communities. While the gap between what ecotourism tends to promise and what it characteristically delivers is evident, even to an armchair analyst, it is not my intention here to examine the political, economic, or social efficacy of this form of travel.[3] Nor do I want to devalue the considerable investment of money, technology and personnel that has gone into developing ecotourism in an attempt to find a way out of the economic malaise and environmental degradation that has been an all-too-common legacy of European imperialism, particularly in the Caribbean and various parts of Africa. Instead, this essay examines the discursive tensions between ecotourism's stated claim to environmental responsibility and its simultaneous imperative to provide predominantly Western clients with an 'authentic' wilderness experience. By reading some of the key visual images and narrative tropes associated with ecotourism alongside their counterparts in colonial discourses such as travel writing, I hope to establish connections that might historicize the current rhetorical purchase of ecotourism as well as provide the basis for an anticolonial critique of the field.

Ecotourism has been variously defined and is at best a slippery term whose modishness has clearly led to a fair amount of indiscriminate application, particularly in some sectors of the tourist industry where ecotourism has come to mean any activity that can be marketed as nature-based. A cursory glance at various literature in the field reveals the ubiquity of descriptors such as *alternative tourism*, *environmental travel*, *green tourism*, *low-impact tourism*, *ethical travel*, and *soft-adventure tourism*, which collectively indicate not only the diversity of practices which have been discussed under the purview

---

[2] In Australia, Canada and New Zealand, for instance, a great number of organized ecotours involve visits to territories primarily populated by Aboriginal, First Nations, and Maori peoples respectively.

[3] For an extended discussion of ecotourism's general failure to deliver its promised solutions to the current crisis of environmental sustainability, see Joe Bandy's provocative survey of the field, "Managing the Other of Nature: Sustainability, Spectacle, and Global Regimes of Capital in Ecotourism," *Public Culture* 8 (1996): 539–66.

of ecotourism but also the industry's concentrated attempts to capture a niche market by selling a (fantasized) dissociation from the rituals of mass tourism. Analysts and industry regulatory bodies adopt much narrower definitions of ecotourism, generally stressing relationships between resource conservation and specific kinds of tourist infrastructure and activity.[4] According to the Ecotourism Society, an international body of tour operators, conservation groups, local communities and host governments, ecotourism is properly defined as "purposeful travel to natural areas to understand the culture and natural history of the environment, taking care not to alter the integrity of the ecosystem, while producing economic opportunities that make the conservation of natural resources beneficial to local people."[5] Central issues in the definitional debate include the degree to which ecotourism encompasses both natural and cultural heritage experiences, and whether certain so-called eco-activities properly belong to the distinctly different genre of adventure travel.[6] In this discussion, I follow the broader usages of the term, while keeping in mind the ecologically based model to which it ideally refers. My commentary pertains generally to organized forms of ecotourism, though this is not to exclude the significant category of 'do-it-yourselfers' likely to be influenced by ecotourism's commercial discourses.

Probably the most consistent thing that ecotourism sells is a first-hand experience of nature, an opportunity to feel, see and appreciate a *natural* – that is, supposedly unaltered – landscape. That many tourist destinations which promise access to this particular eco-experience are located within former European colonies invites a revisiting of some of the existing arguments about imperial constructions of "nature" as an ontological category. The broad field of postcolonial studies has delivered useful and sophisticated accounts of the tensions and contradictions surrounding representations of nature in colonial contexts as, variously, a rich resource to be exploited for the benefit of distant capital interests, a threat to the civilizing march of

[4] Peter S. Valentine, "Ecotourism and Nature Conservation: A Definition with some Recent Developments in Micronesia," in *Ecotourism: Incorporating the Global Classroom*, ed. Betty Weiler (St Lucia & Brisbane: U of Queensland P, 1992): 5; Robert C. Scace, "An Ecotourism Perspective," in *Tourism and Sustainable Development: Monitoring, Planning, Managing*, ed. J.G. Nelson et al. (Waterloo: University of Waterloo, 1993): 64–65.

[5] Megan Epler Wood, Frances A. Gatz & Kreg Lindberg, "The Ecotourism Society: An Action Agenda," in *Ecotourism and Resource Conservation: A Collection of Papers*, comp. Jon A. Kusler (Berne NY: Ecotourism and Resource Conservation Project, 1991): 75.

[6] Scace, "An Ecotourism Perspective," 63.

imperial modernity, a necessary refuge from this very process, and an enigmatic – even capricious – force which may occasionally yield its secrets to the careful naturalist / observer. It is the anodyne version of nature which interests me most here, since its perceived capacity to re-energize the imperial adventurer sapped by the demands of a rapidly industrializing world also speaks to a sense of Western spiritual malaise to which ecotourism more subtly offers a corrective.

Hans Magnus Enzensberger has argued that modern tourism's valorization of nature in its pristine forms dates back to the writings of European Romanticism, which cemented the textualization of the notion that an encounter with forms untainted by human handiwork could provide an (impossible[7]) antidote to the effects of modernity. The Romantic authors, he maintains, "transfigured freedom and removed it into a realm of imagination, until it coagulated into a distant image of a nature far from all civilization, into a folkloric and monumental image of history."[8] That colonial discourse tends to be animated by elegiac and pastoral modes of representing nature is amply demonstrated by the plangent laments of numerous nineteenth-century travel writers engaged in a utopian quest for some sort of Edenic wilderness located in Europe's distant colonies. Moreover, Enzensberger's claim that the sense of a "pristine landscape and untouched history have remained the models of tourism"[9] seems to be more prophetic than he might have imagined when he first published his findings in 1958. More recently, environmentalists such as David Rothenberg have examined the binary relation between hegemonic models of "civilization" and idealized versions of untouched nature as crystallized in the Western concept of "wilderness." Rothenberg insists that "the idea of wilderness has shown itself time and again to be the creation of human consciousness, malleable in the extreme, used to fulfil our deepest desires and worst fears."[10] His contention that wilderness is "an ethnocentric concept that has little to do with

---

[7] Enzensberger sees the quest as impossible because it is caught up in a dialectic of process that means nature's restorative effect is already destroyed at the instant of human contact.

[8] Hans Magnus Enzensberger, "A Theory of Tourism," tr. Gerd Gemünden & Kenn Johnson, *New German Critique* 68.2 (1996): 125.

[9] Enzensberger, "A Theory of Tourism," 125.

[10] David Rothenberg, "Wildness Untamed: The Evolution of an Ideal," Introduction to *Wild Ideas*, ed. Rothenberg (Minneapolis: U of Minnesota P, 1995): xviii.

the more profound and direct ways in which nature is experienced by the world's peoples"[11] raises questions about ecotourism's obsessive interest in wilderness destinations and, indeed, its mandate to contribute to the preservation of wilderness itself.

That wilderness is a highly saleable commodity in the ecotourism business is evident in the plethora of travel brochures that use adjectives such as *pristine*, *isolated*, *remote*, *unspoiled* and so forth, to extol the virtues of their specific destinations, all of which are marketed as unique in an uncannily similar fashion. Such transparent manipulations of wilderness tropes would make it easy to focus a critique of ecotourism on its strategic marketing but, ultimately, that seems a soft target. Perhaps more telling are the multiple contradictions which this discursive harnessing of wilderness tropology reveals. First, the fact that ecotourism brings into the circuit of commodity relations a form of nature – the wilderness – which is, by definition, outside that circuit, supports Enzensberger's view that "the pristine is an ideological mystification" designed to appeal to the modern tourist's sense of nostalgia for a pre-industrial world.[12] In this respect, the very designation of areas as wilderness presumes a prior commodification of natural resources as potential eco-destinations.

Secondly, in a related dialectic, ecotourism in wilderness areas sells an encounter with "unspoiled nature," but one which is structured so that visitors can wilfully ignore the fact that their mere presence is incompatible with the concept being sold, since "unspoiled" in this context implies outside the realm of human activity. (Hence, successful ecotours typically offer low levels of contact with other tour groups.) A recent attempt by the Audubon society to define the infrastructural needs of the ecotourist reveals some of the complexities involved in commodifying the wilderness in a manner that will appeal to the targeted clientele's aesthetic sensibilities and their moral obligation to travel responsibly. The Audubon report warns, for instance, that tourists will be dissatisfied if walking tracks are too rough, but that "care should be taken not to overdevelop the trails [as] ecotourists prefer the conditions to *appear* to be as rugged as possible and to fit the environment."[13] Descriptions of model accommodation, modes of travel and

---

[11] Rothenberg, "Wildness Untamed," xv.

[12] Enzensberger, "A Theory of Tourism," 127.

[13] Ray E. Ashton, "Defining the Ecotourist Based on Site Needs," in *Ecotourism and Resource Conservation: A Collection of Papers*, comp. Jon A. Kusler (Berne NY: Ecotourism

restaurant facilities are framed by similar imperatives to find just the right balance between the comfort of "civilization" and the frisson of a wilderness experience. What this report inadvertently betrays is that ecotourism must distance its clientele from that to which it promises proximity. Hence many travel brochures anxiously stress issues such as security and comfort, usually in the same breath as they proffer a genuine encounter with the wild, the untamed. Wilderness, it seems, is clearly more palatable to the Western consumer in its commodified form, a point incidentally demonstrated by a recent proposal to put up a series of "wildernesses" inside California's shopping malls. These fabricated nature preserves, complete with "wild" flora and fauna, and even the facility for camping, are purportedly designed to fulfil the needs of consumers who "yearn to get back to nature but don't have the time." The experiment will be called (apparently without irony) "The American Wilderness Experience."[14] This postmodern "tendency to blend a nostalgia for an earlier and simpler era with a reassurance that modern conveniences and progress are never far away"[15] resurfaces regularly in ecotourism, as seen in the habit of ending tours with an add-on stay in an adjacent luxury resort. Some seasoned ecotravellers even see such a finale as almost mandatory: "If you've been on a rugged field trip in the tropics [...] do your body and spirit a *big* favour after your tour – check into a deluxe hotel with air-conditioning and a pool."[16]

A third problem is raised by ecotourism's tendency to conceive of the wilderness in terms which exclude routine human activity, an equation which fails to account for the historical presence of indigenous peoples living in "pristine" ecosystems worldwide.[17] Most often, this contradiction is addressed by drawing such peoples into the field of sites on the ecotourist's itinerary. Hence visits to archaeological remains and to "traditional" villages are a popular feature of many ecotours, particularly in developing regions. (It is

and Resource Conservation Project, 1991): 95 (my emphasis).

[14] For a fuller description of the entire proposal, see "The American Wilderness Experience," *Wall Street Journal* (7 August 1997): 23, and http://www.ogdencorp.com/news/ 050296G. htm

[15] Brian E.M. King, *Creating Island Resorts* (London: Routledge, 1997): 214.

[16] Alice Gieffen & Carole Berglie, *Eco Tours and Nature Getaways* (New York: Clarkson, 1993): 15.

[17] R. Edward. Grumbine, "Wise and Sustainable Uses: Revisioning Wilderness," in *Wild Ideas*, ed. David Rothenberg (Minneapolis: U of Minnesota P, 1995): 10.

interesting to note here that Caribbean ecotourism has been most strongly developed in Dominica, Belize and Guyana, countries which all have remnant native Carib populations that are visited as part of each region's standard ecotours.) While the coupling of wilderness with indigenous cultures may accurately reflect some groups' epistemological approaches to nature,[18] it also has the effect of positioning them as objects of a neo-imperial gaze. Like their ancestors, modern-day 'traditional' societies function in many travel-related discourses as primitive Others against which the civilized Self can be defined. The fact that ecotourism replicates mass tourism's interest in this kind of sightseeing confirms the enduring currency of primitivism as a hot commodity whose malleability, like that of the wilderness, always serves the needs of the present. As Marianna Torgovnik has pointed out, "The primitive does what we ask it to do. Voiceless, it lets us speak for it. It is our ventriloquist's dummy – or so we like to think."[19]

If the commodification of pristine forms of nature reveals a fracture at the heart of ecotourism's ideological project, the rhetoric of discovery that goes hand in hand with images of wilderness suggests further points of contact between ecotourism and colonial travel, at least as the two practices have been textualized. Terms such as *expedition*, *exploration* and *odyssey* – frequently used in tour companies' registered names as well as in their descriptions of particular itineraries – are the lexical staples of ecotourism advertising, reminding us of its ideological links with conquest narratives and nineteenth-century travellers' tales. In her work on colonial travel writing in the Caribbean, Claudia Brandenstein examines the ways in which Charles Kingsley, for instance, casts himself in the roles of discoverer and explorer, modelling his travels on those of historical figures such as Columbus.[20] Similarly, the modern ecotourist is characteristically offered the "opportunity of a lifetime" to discover – apparently for the first time – various unique and spectacular features of a remote region. In some ecotravel literature, explicit reference to European explorers and/or missionaries (notably Columbus

---

[18] This position tends to be commensurate with deep ecology's rethinking of the nature-culture binary along the lines of various indigenous philosophies that do not draw ontological distinctions between the two.

[19] Marianna Torgovnik, *Gone Primitive: Savage Intellects, Modern Lives* (Chicago: U of Chicago P, 1990): 8.

[20] Claudia Brandenstein, "Imperial Positions in Charles Kingsley's *At Last: A Christmas in the West Indies*," *SPAN* 46 (1998): 6–7.

for Caribbean destinations and Livingstone for the typical African eco-safari) positions the contemporary traveller's act of discovery not merely as a mode of learning about an environment that is new to the individual but rather as a way of actually participating in an ongoing historical endeavour. Thus the ecotourist implicitly extends the great imperial voyages and treks of discovery. But if the main object of colonial exploration was to identify potential resources for the expansion of the Empire's "great estate," the end-point of eco-discovery is more personal and potentially much more ethical: self-discovery. This emphasis on self-discovery aligns ecotourism with adventure travel and indeed the latter is often featured as a subset of the former.[21] Where adventure travel differs markedly from the purer forms of ecotourism is in its philosophical attitude to the environment: the true ecotourist seeks wilderness in order to commune with nature rather than to master it.

If the concept of wilderness and its associated rhetoric indicates one connecting point between ecotourism and colonial travel, the shared interest in learning points to another. In industry definitions as well as in marketing material, ecotourism stresses the potential of travel as an epistemological mode. A 1991 Queensland symposium titled "Ecotourism: Incorporating the Global Classroom" supports this notion in its very nomenclature and more than one commentator has drawn on the classroom metaphor to explain the links between ecotourism's experiential focus and its presumed educational value. An old Chinese aphorism apparently says it all: "I hear and I forget; I see and I remember; I do and I understand."[22] Policy and planning documents in the field also argue that ecotourism's hands-on approach has high cognitive dimensions and that tourist gratification is measured largely in terms of education. In simple terms, then, the ecotourist wants and *needs* to learn, and it is the function of tour operators and host communities to provide ample opportunities for that to happen. Sally Grotta's naive assessment of the motivations behind standard eco-activities nonetheless emphasizes the perceived strength of the ecotourist's gnostic drive:

---

[21] According to Tourism Canada, adventure travellers "expect to experience varying degrees of risk, excitement and tranquillity and to be personally tested or stretched in some way. They are explorers of both an outer world, the unspoiled exotic parts of our planet, and an inner world of personal challenge, self-perception, and mastery"; quoted in Scace, "An Ecotourism Perspective," 64.

[22] Cf. Roger Grant & Terry O'Brian, "Ecotourism – Educational Travel: A Growth Business," in *Ecotourism Business in the Pacific: Promoting a Sustainable Experience*, ed. John E. Hay (Auckland: Environmental Science Occasional Publications, 1992): 71.

> the ecotourist is also the most intelligent and the most caring of tourists. The fact that he or she has chosen a tour that visits isolated mountain villages or a cruise that explores small, undeveloped islands indicates his [sic] disposition and inclination to learn. [...] Ecotourists don't just travel to have a good time, but to have a good time by learning.[23]

While valorization of the quest for knowledge about the environment stems partly from ecotourism's early grassroots connections with organizations such as Earthwatch and the Smithsonian Institution, the central tenets of the travel-learn concept can once again be traced to key features of specific subgenres of the colonial travelogue. This is not an incidental comparison but rather one which illustrates the Western imaginary's continued investment in a view of nature guided by Enlightenment forms of rationality.[24] Indeed, Gieffen and Berglie introduce their 1993 guide by arguing that an ecotour is "a naturalist's expedition in twentieth-century terms."[25]

In her study of travel writing and imperial expansion in the eighteenth and nineteenth centuries, Mary Louise Pratt identifies the naturalist as a paradigmatic figure whose will to learn via travel to "exotic" locations to catalogue specimens established a particular eurocentric world-view, thereby naturalizing the bourgeois European's own global presence and authority:

> One by one, the planet's life forms were to be drawn out of the tangled threads of their life surroundings and rewoven into European-based patterns of global unity and order. The (lettered, male, European) eye that held the system could familiarize ("naturalize") new sites/sights immediately upon contact, by incorporating them into the language of the system.[...] Natural history extracted specimens not only from their organic or ecological relations with each other, but also from their places in other peoples' economies, histories, social and symbolic systems.[26]

---

[23] Sally Wiener Grotta, "The Ecotourist as Ambassador," in *Ecotourism and Resource Conservation: A Collection of Papers*, comp. Jon A. Kusler (Berne NY: Ecotourism and Resource Conservation Project, 1991): 102.

[24] Andrew Dobson argues that the spread of Enlightenment rationality underpins our exploitative relationship with the natural world. From this (Baconian) point of view, "nature has no meaning in itself; rather its meaning comes from our instrumental apprehension of it"; Dobson, "Critical Theory and Green Politics," in *The Politics of Nature: Explorations in Green Political Theory*, ed. Andrew Dobson & Paul Lucardie (London: Routledge, 1993): 193.

[25] Gieffen & Berglie, *Eco Tours and Nature Getaways*, 2.

[26] *Imperial Eyes: Travel Writing and Transculturation* (London: Routledge, 1992): 31.

The particular power of natural history travelogues, Pratt argues, stemmed in part from the "conspicuous innocence of the naturalist," an innocence constituted "in relation to the presumed guilt of conquest."[27] Beside the seafarer or the conqueror, the naturalist appeared decidedly benign, interested only in a non-exploitative relationship with nature via the scientific classification of species.

In a somewhat similar fashion, the presumed neutrality of scientific inquiry has functioned in Third-World destinations as an 'anti-conquest' narrative that sets the modern ecotourist apart from the implicitly neo-imperial mass tourist. In the Caribbean, for example, an Earthwatch tour to San Salvador involves "monitoring of oceanic pollution" during a visit to local seagrass meadows, while the American Oceanic Society offers a "research swim with dolphins" as part of its Bahamas expedition, and the London-based Field Studies Council conducts "botanizing" trips to remote mountain regions of Jamaica.[28] Such scientific endeavours transform idle tourist pleasures such as snorkelling, swimming and wildlife-viewing into purposeful "work" that speaks to the ecological imperative for everybody to do his or her bit to "save the planet." If the proportion of the ecotourism market originally served by these researched-based organizations has shrunk due to the rapid expansion of wholly commercial ecotour companies, the spirit of scientific inquiry is nonetheless harnessed to sell a range of contemporary eco-destinations, even though research activities have morphed into other forms of environmental education that suppose ecological outcomes to follow naturally from an individual's travel-learn experience. Both kinds of ecotourism tend to see knowledge acquisition as an important way of preventing the negative environmental effects of a modern industrialized and technologized world. But, as a variety of literature in the field shows, the division between instrumentalist and ecological uses of nature is not so easily maintained, even in the discursive realm, much less in actual practice. Just as Pratt's naturalist eternally invokes the guilt of conquest by trying to distance "himself" from imperial exploitation,[29] the ecotourist comes into being as an ontological category only within the broader referential frame of mass tourism's apparent environmental ignorance and irresponsibility.

---

[27] Pratt, *Imperial Eyes*, 57.

[28] Gieffen & Berglie, *Eco Tours and Nature Getaways*, 91–93.

[29] Pratt, *Imperial Eyes*, 57.

Studies in colonial discourse show that naturalists produced commercially exploitable knowledge and were therefore implicated, wittingly or unwittingly, in the march of imperial capitalism. Ecotourists, on the other hand, are generally thought of as post-imperial subjects, ethical travellers whose nature-based activities actually advance conservation efforts and benefit local economies. Whether or not this ideal is achieved, it may be instructive to consider that ecotourism to 'underdeveloped' regions replicates natural history insofar as it produces a Western discourse about non-Western worlds, "an urban discourse about non-urban worlds, and a lettered, bourgeois discourse about non-lettered, peasant worlds."[30] An important difference, however, is that natural history democratized science primarily at the discursive level by taking readers on a vicarious journey of discovery, whereas ecotourism potentially offers everyone the chance to *experience* nature in a quasi-scientific way. Indeed, a common argument about the environmental efficacy of ecotourism is that such detailed encounters with nature enable tourists to develop emotional attachments that will automatically lead to positive conservation outcomes.[31] Implicit in this equation is the idea that specific sites belong to everyone who is willing to study their ecosystems first-hand, that such sites are somehow separable from local cultures and land use patterns. Since it is typically the Western consumer who has the greatest opportunity to learn in this way, designated eco-sites in all areas of the globe tend to circulate within the parameters of both Western discourse and Western commerce.[32] A number of critics warn about the environmental perils of this situation, arguing that ecotourism represents the thin edge of the wedge to mass tourism because it brings highly sensitive physical and cultural environ-

---

[30] Pratt, *Imperial Eyes*, 34–35.

[31] See Gieffen & Berglie, *Eco Tours and Nature Getaways*, 2; Hector Ceballos–Lascurain, "Tourism, Ecotourism and Protected Areas," *Parks* 2.3 (1991): 31–32; Grotta, "The Ecotourist as Ambassador," 107. King argues that detailed 'scientific' description also functioned in colonial times as an important step in the process of settlement; to survey the land in this manner led to a sense of attachment that justified disenfranchising indigenous inhabitants; *Creating Island Resorts*, 207.

[32] Brenda Rudkin and C. Michael Hall note that the driving force for ecotourism in developing countries comes from "foreign donors, investors, academic institutions, consultants and conservation groups"; Rudkin & Hall, "Unable to See the Forest for the Trees: Ecotourism Development in the Solomon Islands," in *Tourism and Indigenous Peoples*, ed. Richard Butler & Thomas Hinch (London: International Thomson Business Press, 1996): 223.

ments to the notice of a broader international market,[33] a claim that seems to have been borne out in a number of cases. In Belize, for instance, conservationists and government bodies are now grappling with a significant increase in overall tourism numbers since the massive reef that lies just off the country's coastline became a popular eco-destination.[34]

The pedagogical imperative that underpins most definitions of ecotourism is brought to bear on a number of debates within the field, including the question of whether so-called "consumptive" uses of nature constitute legitimate eco-activities. Consumptive activities include hunting and fishing, while non-consumptive practices focus on wildlife viewing, bird watching, photography and so forth. In general, the latter are deemed primarily educational (and aesthetic), the former potentially exploitative, though various agitators insist that controlled hunting and fishing should fall within ecotourism's ambit. Interestingly, Ray Ashton calls on the tradition of natural history expeditions to validate this view, maintaining that ecotourism is "not a new endeavour" but merely "a new term" to describe what (Western) people have been doing since the 1800s: "travel[ing] on safaris to exotic parts of the world [...] to hunt or collect or just to experience the exotic wildlife, jungles and cultures"; in the same breath as he points to trophy-hunting as a continued area of interest, Ashton argues that today's ecotourists are distinguished by their willingness to settle for "great photographs, a full bird checklist or just fond memories."[35] While most other commentators situate ecotourism as a thoroughly contemporary development designed to respond to environmental concerns, Ashton's genealogy of the field, albeit simplistic in the extreme, once again raises questions about Western culture's historical uses of rare or exotic species.

In a fascinating account of nineteenth-century British zoological exhibitions and acclimatization programmes, Harriet Ritvo demonstrates that exotic animals were important sites of conquest in the imperial campaign: "the maintenance and study of captive wild animals, simultaneous emblems

---

[33] See Erlet Cater, "Ecotourism: A Sustainable Option?" *Geographical Journal* 159.5 (1993): 114–15.

[34] This problem is documented in the educational video *Cashing in on Paradise*, written and produced by Nicola Ebenau, Geographical Society Films, 1993. Polly Patullo lists other Caribbean examples of environmental damage resulting from ecotourism; Patullo, *Last Resorts: The Cost of Tourism in the Caribbean* (London: Cassell, 1996): 104–35.

[35] Ray Ashton, "Defining the Ecotourist Based on Site Needs," 91.

of human mastery over the natural world and of English dominion over remote territories, offered an especially vivid rhetorical means of reenacting and extending the work of empire."[36] Zoos and menageries, Ritvo argues, metonymically displayed the magnitude of the British Empire, inviting the public to share figuratively in the Empire's wealth as well as in its scientific achievement in capturing, transporting and keeping alive wild animals (209–10). Such collections were also powerful "reminders of the hunting expeditions during which [the animals] had been procured" (247), and so functioned to sustain the master-narrative of confrontation and conquest that lay at the heart of European expansionism. Ritvo, among others, points out that these hunting exhibitions supported the march of Empire in the discursive as well as the literal realm through the widespread dissemination of sporting tales and adventure narratives that allowed "even humble citizens to engage, at least by proxy, in a kind of metaphoric reenactment of the conquest that had previously been assigned to the privileged classes" (257).

I would argue that ecotourism's pervasive interest in "wildlife experience" shares a number of important features with colonial uses of exotic animals, employing a similar instrumentalist agenda that values species according to their ability to give humans aesthetic pleasure. Just about all ecotours feature some kind of encounter with wildlife and many are structured around this specific experience – if only as a possibility. Planning documents in the field suggest that consumer satisfaction is directly proportional to the volume of wildlife experienced, with one analyst even specifying that, "as a general rule, each day should have one or two outstanding wildlife experiences that one can count on seeing at least 60% of the time" plus other interesting scenic points or wildlife encounters along the way.[37] While much of the relevant discourse is decidedly vague about just what constitutes a "wildlife experience," it is clear that this is definitely not the same thing as merely viewing rare species in their natural habitats. What appears to be the fundamental difference is the way in which the ecotourist's encounters with nature are structured: almost invariably as some kind of quest that eventually yields its elusive prize. Just as colonial accounts of big-game hunting or ventures of that ilk "emphasized the difficulties and dangers encountered by the hunter

---

36 Harriet Ritvo, *The Animal Estate: The English and Other Creatures in the Victorian Age* (Cambridge MA: Harvard UP, 1987): 205. Further page references to Ritvo in this paragraph are in the main text.

37 Ashton, "Defining the Ecotourist Based on Site Needs," 93–94.

in order to magnify his eventual triumph,"[38] the contemporary ecotourist expects – indeed demands – a wildlife experience that tells of personal challenges and triumphs. Hence many tours entail physically demanding journeys to animal, bird or fish habitats, along with an (often implied) promise of close contact with potentially dangerous wildlife. Native guides are often used, as they were in colonial times, to ensure a successful "hunt." It is their task to make available their specialized knowledges about particular species, and, in some cases, to facilitate a quasi-spiritual encounter between tourists and their "prey."

These parallels suggest that ecotourism's typical "wildlife experience" may operate within the same discursive economy as the colonial hunting safari, in so far as both excursions turn on the (Western) traveller's heroic confrontation with species symbolic of uncivilized nature. Of course, the ecotourist's material proof of the encounter – a full checklist, stunning photos, or perhaps a first-hand sketch – differs greatly from the big-game hunter's booty – usually some part of the animal itself – yet both constitute souvenirs in Susan Stewart's sense of fetishized objects that tell a narrative of acquisition rather than revealing details of the actual thing represented. According to Stewart, "to have a souvenir of the exotic is to possess both a specimen and a trophy; on the one hand the object must be marked as exterior and foreign; on the other, it must be marked as arising directly out of an immediate experience of its possessor."[39] In essence, then, the souveniring process requires the textualization of objects, usually through pictorial and/or written records such as might constitute the typical ecotourist's daily log. That colonial hunters tended to keep journals which recorded game encounters, near misses, game shot, and comments about weather and environment[40] illustrates that this kind of inscription has a long history.

By drawing analogies between ecotourism and big game hunting, I do not wish to argue that activities such as photographing animals, despite the suggestive idioms of "loading" and "shooting" the camera, amount to the same thing as actually killing them. Clearly, there is a world of difference between the moral attitudes of the hunter and the ecotourist, as well as between the

---

[38] Ritvo, *The Animal Estate*, 257.

[39] Susan Stewart, *On Longing: Narratives of the Miniature, the Gigantic, the Souvenir, the Collection* (Baltimore MD: Johns Hopkins UP, 1984). 147.

[40] John M. MacKenzie, *The Empire of Nature: Hunting, Conservation and British Humanism* (Manchester: Manchester UP, 1988): 34.

environmental consequences of their respective actions. Instead, my account of ecotourism's structured "wildlife experience" is intended to show that the commodification of natural species, particularly those considered rare or foreign, continues to underpin Western approaches to the environment. One might even argue that ecotourism effectively democratizes the big-game hunt as much as it repositions natural history within the realm of the common traveller. In this respect, the differences between consumptive and non-consumptive uses of wildlife *as they are currently understood* may not be so great, a situation which possibly explains the minority view of hunting and fishing as legitimate eco-activities.

The marketing emphasis for Third-World eco-destinations rests not just on wildlife experience but also on "authentic" encounters with indigenous peoples, often "packaged" in terms of what Valene Smith has discussed as "the four Hs": habitat, heritage, history, and handicrafts.[41] In brief, the rhetoric of authenticity demands that such peoples are traditional rather than westernized, that encounters are spontaneous rather than staged, and that the tourist attains some degree of intimacy with the "natives." Grotta's comment that "this intimate relationship with other people in other places used to be the private domain of explorers, anthropologists [and] missionaries"[42] suggests a lingering nostalgia for the European colonial realm and situates ecotourism, once again, within its discursive reach. The contradictions intrinsic to ecotourism's general take on authenticity are revealed incidentally by Grotta's insistence that locals must be consulted about "*organising* personal encounters" with ecotourists even though they "may not be as sophisticated as the tour provider" in such matters.[43] Apparently, "the ecotourist won't mind relative discomfort or surprises in his surroundings if he understands that the experience is authentic."[44] This insistence on a "non-touristy" interaction betrays the industry's deep anxiety about the very impossibility of an authentic experience. As Frow has demonstrated, all tourism turns on a fundamental paradox because the tourist experience is always already structured so that it destroys the authenticity of the desired Other by bringing

---

41 Valene L. Smith, "Indigenous Tourism: The Four Hs," in *Tourism and Indigenous Peoples*, ed. Richard Butler & Thomas Hinch (London: International Thomson Business Press, 1996): 283–307.

42 Grotta, "The Ecotourist as Ambassador," 101.

43 "The Ecotourist as Ambassador," 104 (emphasis added).

44 "The Ecotourist as Ambassador," 104.

that Other into the circuit of commodity relations, while nevertheless compelling a "real" encounter between the tourist and the object of interest.[45]

If the real is in fact unattainable in this context, it seems that successful ecotourism boils down to what Judith Adler might call "the style of travel performance": the particular manner in which one artfully *performs* a journey and its attendant activities.[46] The following passage from a recent brochure by Wildland Adventures, winners of an Ecotourism 1995 Award for their operations in Belize and Honduras, eloquently outlines the ways in which the industry attempts to meet its market's demand for the specific travel style that is ecotourism's stock-in-trade:

> In every destination we work with extraordinary English-speaking indigenous guides who share their personal insights and knowledge of natural and cultural history. They know the offbeat routes, where to find wildlife and how to create natural, authentic and intimate cross-cultural encounters. Our casual and non-intrusive style of travel induces friendship with local residents and camaraderie among fellow travellers. We prefer the most charming, local style accommodations that offer security and comfort without ostentation. (12)

That this kind of venture has sometimes been termed "chic" travel further suggests the centrality of a certain ecotourism-specific style, seen as "avant garde, culturally sensitive and ecologically responsible."[47] Although such a style responds to specific contemporary political, cultural and environmental forces, the broad concept of artfully performing a journey in "exotic" regions of the globe dates back at least to the eighteenth century. What the ecotourism travel style apparently shares with the different modes of colonial travel is the invention of a distinct "travelling persona" whose embodied passage through remote territories gives on to various forms of self-inscription or self-styling, whether as explorer, naturalist, adventurer, or simply hapless onlooker of Empire. But whereas these paradigmatic colonial figures necessarily completed the inscription process in travelogues and various other writings, it seems enough for the modern ecotourist to simply display his or her style through travel itself. At the same time, the trans-generic impulse of colonial travel writing to assert discursive control over various sites / sights of conquest becomes ecotourism's present-day emphasis on

[45] John Frow, "Tourism and the Semiotics of Nostalgia," 146.

[46] Judith Adler, "Origins of Sightseeing," *Annals of Tourism Research* 16 (1989): 24.

[47] Ira Silver, "Marketing Authenticity in Third World Countries," *Annals of Tourism Research* 20 (1993): 315.

self-control, both in terms of developing intellectual, physical and moral competencies to respond to environmental challenges, and by avoiding the conspicuous consumption of mass tourism.

Richard Grove insists that modern global environmental consciousness has complex and yet identifiable roots "in the encounters of a whole variety of innovative thinkers with the drastic ecological consequences of colonial rule and capitalist expansion."[48] It follows, then, that any serious examination of ecotourism as a textual field should factor in the relevant historical data on colonial approaches to the environment, though this is by no means the norm in recent cultural criticism.[49] While I have focused on significant points of contact between past and present styles of travel in colonized regions in order to outline a conceptual trajectory that warns of ecotourism's pitfalls, further analysis might reveal equally significant differences. We might ask, for instance, how the ecotourist's phenomenological appreciation of the landscape departs from established nineteenth-century aesthetic conventions such as the picturesque and the panoramic. The fact that the majority of customers on organized ecotours are women whereas the grand colonial tour was normally (though not exclusively) the precinct of men also raises issues worth investigating. My more modest project has been to problematize facile accounts of ecotourism as a new and entirely ethical practice. Far from signalling a major paradigm shift in Western attitudes to non-Western environments, ecotourism seems to be the current buzzword for a cluster of related practices and discourses that too often replay a Romantic nostalgia for other, less industrialized, worlds. In this respect, it is perhaps less philosophically useful to delineate responsible from questionable forms of ecotourism than to remember that neither can be a panacea to the global environmental stress initiated by imperial modernity.

---

[48] Richard H. Grove, *Green Imperialism: Colonial Expansion, Tropical Island Edens and the Origins of Environmentalism 1600–1860* (Cambridge: Cambridge UP, 1995): 474.

[49] A notable exception is Patrick Holland and Graham Huggan's historicized treatment of "ecotopias" in contemporary travel writing; Holland & Huggan, *Tourists with Typewriters: Critical Reflections on Contemporary Travel Writing* (Ann Arbor: U of Michigan P, 1998): 178–95.

WORKS CITED

Adler, Judith. "Origins of Sightseeing," *Annals of Tourism Research* 16 (1989): 7–29.

"The American Wilderness Experience," *Wall Street Journal* (7 August 1997): 23.

Ashton, Ray E. "Defining the Ecotourist Based on Site Needs," in *Ecotourism and Resource Conservation*, com. Kusler, 91–98.

Bandy, Joe. "Managing the Other of Nature: Sustainability, Spectacle, and Global Regimes of Capital in Ecotourism," *Public Culture* 8 (1996): 539–66.

Brandenstein, Claudia. "Imperial Positions in Charles Kingsley's *At Last: A Christmas in the West Indies*," *SPAN* 46 (1998): 4–18.

Butler, Richard, & Thomas Hinch, ed. *Tourism and Indigenous Peoples* (London: International Thomson Business Press, 1996).

Cashing in on Paradise. Written & prod. Nicola Ebenau. Geographical Society Films, 1993.

Cater, Erlet. "Ecotourism: A Sustainable Option?," *Geographical Journal* 159.5 (1993): 114–15.

Ceballos–Lascurain, Hector. "Tourism, Ecotourism and Protected Areas," *Parks* 2.3 (1991): 31–35.

Dobson, Andrew. "Critical Theory and Green Politics," in *The Politics of Nature: Explorations in Green Political Theory*, ed. Andrew Dobson & Paul Lucardie (London: Routledge, 1993): 190–209.

Enzensberger, Hans Magnus. "A Theory of Tourism," tr. Gerd Gemünden & Kenn Johnson, *New German Critique* 68.2 (1996): 113–35.

Frow, John. "Tourism and the Semiotics of Nostalgia," *October* 57 (1990): 123–51.

Gieffen, Alice, & Carole Berglie. *Eco Tours and Nature Getaways* (New York: Clarkson, 1993).

Grant, Roger, & Terry O'Brian. "Ecotourism – Educational Travel: A Growth Business," in *Ecotourism Business in the Pacific: Promoting a Sustainable Experience*, ed. John E. Hay (Auckland: Environmental Science Occasional Publications, 1992): 71–74.

Grotta, Sally Wiener. "The Ecotourist as Ambassador," in *Ecotourism and Resource Conservation*, com. Kusler, 99–108.

Grove, Richard H. *Green Imperialism: Colonial Expansion, Tropical Island Edens and the Origins of Environmentalism 1600–1860* (Cambridge: Cambridge UP, 1995).

Grumbine, R. Edward. "Wise and Sustainable Uses: Revisioning Wilderness," in *Wild Ideas*, ed. David Rothenberg (Minneapolis: U of Minnesota P, 1995): 3–25.

Holland, Patrick, & Graham Huggan. *Tourists with Typewriters: Critical Reflections on Contemporary Travel Writing* (Ann Arbor: U of Michigan P, 1998).

King, Brian E.M. *Creating Island Resorts* (London: Routledge, 1997).

Kusler, Jon A., comp. *Ecotourism and Resource Conservation: A Collection of Papers* (Berne NY: Ecotourism and Resource Conservation Project, 1991).

MacKenzie, John M. *The Empire of Nature: Hunting, Conservation and British Humanism* (Manchester: Manchester UP, 1988).

Patullo, Polly. *Last Resorts: The Cost of Tourism in the Caribbean* (London: Cassell, 1996).

Pratt, Mary Louise. *Imperial Eyes: Travel Writing and Transculturation* (London: Routledge, 1992).

Ritvo, Harriet. *The Animal Estate: The English and Other Creatures in the Victorian Age* (Cambridge MA: Harvard UP, 1987).

Rothenberg, David. "Wildness Untamed: The Evolution of an Ideal," Introduction to *Wild Ideas*, ed. David Rothenberg (Minneapolis: U of Minnesota P, 1995): xii–xxvii.

Rudkin, Brenda, & C. Michael Hall. "Unable to See the Forest for the Trees: Ecotourism Development in the Solomon Islands," in *Tourism and Indigenous Peoples*, ed. Butler & Hinch, 203–26.

Scace, Robert C. "An Ecotourism Perspective," in *Tourism and Sustainable Development: Monitoring, Planning, Managing*, ed. J.G. Nelson et al. (Waterloo, Ontario: University of Waterloo, 1993): 59–82.

Silver, Ira. "Marketing Authenticity in Third World Countries," *Annals of Tourism Research* 20 (1993): 302–18.

Smith, Valene L. "Indigenous Tourism: The Four Hs," in *Tourism and Indigenous Peoples*, ed. Butler & Hinch, 283–307.

Stewart, Susan. *On Longing: Narratives of the Miniature, the Gigantic, the Souvenir, the Collection* (Baltimore MD: Johns Hopkins UP, 1984).

Torgovnik, Marianna. *Gone Primitive: Savage Intellects, Modern Lives* (Chicago: U of Chicago P, 1990).

Valentine, Peter S. "Ecotourism and Nature Conservation: A Definition with some Recent Developments in Micronesia," in *Ecotourism: Incorporating the Global Classroom*, ed. Betty Weiler (St Lucia & Brisbane: U of Queensland P, 1992): 4–9.

Wood, Megan Epler, Frances A. Gatz & Kreg Lindberg. "The Ecotourism Society: An Action Agenda," in *Ecotourism and Resource Conservation*, com. Kusler, 75–79.

Colonial Nature-Inscription
On Haunted Landscapes

ANDREW McCANN

A PROJECT ENTITLED "ENVIRONMENT AND EMPIRE" implies, if not a paradigm shift, at least a revised sense of what colonization as a process entails. In general we understand colonialism in anthropocentric terms, that is as the violence inflicted by one society (by one group of humans) on another. In this paradigm decolonization is imagined, perhaps provisionally, as the emergence of resistant national collectives, while representations of the non-human environment are easily incorporated into readings of competing ideological structures such that "nature" is seldom discussed in its own right. As Kate Soper writes, this approach tends to focus on the "semiotics of nature," which emphasizes "the role of the concept in mediating access to the 'reality' it names, and whose political critique is directed at the oppressive use of the idea to legitimate social and sexual hierarchies and cultural norms."[1] If the recent emergence of the environment as an urgent ethical and political concern has forced us to rethink this, it has not helped us to extricate the issue of nature from its fatal imbrication with the ideology of empire. If nature – inclusive of the human and the non-human – is the real object of colonization in the sense of instrumental appropriation, as critical theorists like Max

[1] Kate Soper, *What is Nature? Culture, Politics and the Non-Human* (Oxford: Blackwell, 1995): 3.

Horkheimer and Theodor Adorno have implied, then surely some sense of nature as thing-in-itself is a necessary corrective to a world-view that insists on seeing nature as a resource base.[2] And yet it is exactly this view of nature that has been assimilated into nationalisms that are reactionary and militaristic in their commitment to their own blood and soil. Simon Schama's account of the relationship between German nationalism and the mythic memory of the German forest reminds us that ecological consciousness does not necessarily obviate barbarism).[3] In fact, the confluence of ecological awareness and nationalism is a fairly typical feature of the cultural politics of Romanticism. Although a world away from fascist appropriations of *Kulturlandschaft*, Wordsworth's late poem *The Excursion*, for example, has rightly been read as a document of cultural nationalism and a defence of British imperialism in which Romantic depictions of landscape function as a way of evoking a secure and benign sense of British identity. Yet the poem also includes laments for the fate of nature in the age of industrialization that resonate with a contemporary sense of environmental awareness. Wordsworth, for example, has one of his characters declare:

> I grieve, when on the darker side
> Of this great change I look; and there behold
> Such outrage done to nature as compels
> The indignant power to justify herself;
> Yea, to avenge her violated rights.[4]

It is not satisfying to say that this moment is simply ideological, even though nature and nation clearly evoke each other in this poem. At some very basic, intuitive level I want to agree with what is implied here: that a position of sympathy with nature is both aesthetically and morally normative. This is not to indulge a naive faith in the restorative power of Romanticism, but to suggest that the idea of nature as object and site of lament, as an occasion for elegy that can be disentangled from the idea of nation, is simply Roman-

---

[2] See Andrew Dobson's "Critical Theory and Green Politics" in *The Politics of Nature: Exploration in Green Political Theory*, ed. Andrew Dobson & Paul Lucardie (London: Routledge, 1993): 190–209, for a discussion of early Frankfurt School theory in relationship to more recent ecopolitics.

[3] See Simon Schama, *Landscape and Memory* (London: Fontana, 1995): 75–134, on the cultural politics of the *Holzweg*.

[4] William Wordsworth, "The Excursion" (1814), in *The Poems*, ed. John O. Hayden (Penguin English Poets; Harmondsworth: Penguin, 1977), vol. 2: 255.

ticism's normative moment – a moment at which an aesthetic detachment from the increasingly pervasive imperatives of instrumental rationality makes it possible to glimpse a relationship to nature through which the true malevolence of modernity is evident. Of course, the very possibility of aesthetic detachment needs to be considered dialectically: it is inseparable from the process of social differentiation through which instrumental rationality is itself able to establish its power over nature.

The normative moment referred to above is Romanticism's legacy to critical theory, evident in the latter's frequently elegiac character. In this essay I want to suggest some of the ways in which the refusal of instrumentality in Romantic elegy is compromised by the complicit forces of colonization and commodification, such that a notion of nature as object of lament gives way to representations of landscape in which resistance to instrumentality becomes increasingly difficult to express. This mutation at the level of representation is the result of a tension between the ideally disinterested realm of aesthetic rationality and the social forces that impinge upon it, overwriting it in terms of other imperatives. By looking briefly at the ways in which the conventions of Romantic elegy were adapted by nineteenth-century Australian Romantics, I want to sketch one historically specific example of how the normative impulse that circulates around our need for nature became incorporated into representational practices beholden to instrumentality. In so doing, I want to stress that while representations of nature are obviously cultural achievements, we can still talk about the ethical implications of different representational practices, and thus at least imagine the possibility of a semiotics of nature adequate to a progressive, ecologically conscious politics.

To talk about the political or ethical valences of the aesthetic in this way evokes, for academics working under the aegis of postcolonial theory and/or cultural studies, a cultural imperialism that involves the prioritization of European aesthetic norms and, in the form of high-school and university curricula, their integration into disciplinary practices that have targeted potentially dissident populations in both Europe and its colonies. This is the point of Ian Hunter's critique of nineteenth-century aesthetics as a practice of the self, and accounts for the subsequent abandonment of the category of the aesthetic by the cultural policy movement.[5] But the ways in which

---

[5] Ian Hunter, "Aesthetics and Cultural Studies," in *Cultural Studies*, ed Lawrence Grossberg, Cary Nelson & Paula Treichler (London: Routledge, 1992): 347–72. For a compre-

aesthetic concerns activate political ones are, of course, much more diverse than this position allows. For example, it is now virtually impossible to ignore the ways in which an elegiac idiom, in other words an aesthetic paradigm, organizes a range of oppositional positions in which mourning has become something like a mode of deportment, even a style of citizenship capable of expressing resistance to the ongoing appropriation and brutalization of nature. We see this in the performance of lament as a way of disidentifying with Australia's colonial past and its capitalist present. That this elegy is so obviously political also alerts us to the imbrication of the political and the aesthetic, yet in a way that compels us to reconsider the normative potential of the latter. Indeed the problem of finding a language appropriate to the possibility of environmental catastrophe and sensitive to the ways in which images of nature are ideological is such an urgent one that revisiting the normative moment in a cultural paradigm committed to nature as a locus of aesthetic and moral value might not be as regressive as Romanticism's critics would argue.[6]

The relationship between critical consciousness and aesthetic forms is the object of Adorno's writings on art, music and literature, and ultimately the most important aspect of his contribution to critical theory. In these writings, elegy is not only crucial to the ability of culture to register and bear witness to the violence done to nature by capitalist modernity, it is the crystallization of an aesthetic rationality conceptualized as an *a priori* refusal of the instrumental rationality that is, for critical theorists, the dark side of the enlightenment. Like Romantic elegy, with which his work shares a great deal, Adorno imagines a utopian point of reconciliation beyond the violence of modernity – what Albrecht Wellmer aptly calls the "non-violent unity of the diverse in the reconciliation of all living things."[7] Yet at the same time a good deal of what drives Adorno's work is the need to mark the ways in which radical art refuses the role otherwise ascribed to the aesthetic in the consolidation of bourgeois subjectivity, privacy and pleasure, where it is

---

hensive account of cultural policy, see Tony Bennett, *Culture: a Reformer's Science* (Sydney: Allen & Unwin, 1998).

[6] It is worthwhile considering, moreover, how the current critique of "Romantic" nostalgia might itself offer, as Kate Soper writes, "an extremely partial, and in its own way deluded, account of what it is to be alive in the present"; Soper, *What is Nature? Culture, Politics and the Non-Human* (Oxford: Blackwell, 1995): 201.

[7] Albrecht Wellmer, *The Persistence of Modernity: Essays on Aesthetics, Ethics and the Postmodern*, tr. David Midgley (Cambridge MA: MIT Press, 1991): 5.

tantamount to a form of escapism. Integral to this is art's refusal of a naïve utopianism and the enjoyment it is supposed to yield. If artworks tend *a priori* towards the utopianism that consists in their refusal of the empirical and the instrumental, a utopianism that is an inevitable consequence of aesthetic synthesis, then the fact that their promise of happiness is so easily equated with the pleasures of escapist consumption, integrated into economies of cultural prestige or linked to notions of disciplined subjectivity, such as those discussed by Hunter, means that art, in order to remain radical, must turn against itself, undermining its own promise of unmediated pleasure or wholeness. As Adorno states, "the clichés of art's reconciling glow enfolding the world are repugnant not only because they parody the emphatic concept of art with its bourgeois version and class it among those Sunday institutions that offer solace. These clichés rub against the wound that art itself bears."[8] The critical or normative dimension of art, for Adorno, seems to reside in this notion of the wound, indicating the extent to which, despite its claims to autonomy, art, or the concept of art, also refuses the bourgeois vision of a falsely reconciled world – a vision that is really now a function of lifestyle-oriented advertising, and the leisure and entertainment industries which hold out the promise of undiluted pleasure. The art that Adorno endorses, on the contrary, presents pleasure in terms of a kind of temporal displacement, such that the moment of reconciliation is never actually present. It appears as something apparitional, an afterimage that is discernible as a kind of trace in which the artwork alludes to something beyond its own aesthetic immanence. It is this spectrality that perhaps defines a critical aesthetic: "however powerful, historically, the force of pleasure to return may be, whenever it appears in art literally, undefracted, it has an infantile quality. Only in memory and longing, not as copy or as an immediate effect, is pleasure absorbed by art."[9] Art's commitment to the idea of reconciliation, in other words, is realized in its "renunciation of the semblance of reconciliation."[10] This is a key to Adorno's thinking, and it points to the intent behind the pessimism that defines his writing. While the experience of the Holocaust is central to this aspect of Adorno's work, the elegiac idiom also bears witness to a more general dynamic of which the Holocaust

---

[8] Theodor W. Adorno, *Aesthetic Theory*, tr. Robert Hullot–Kentor (Minneapolis: U of Minnesota P, 1997): 2.

[9] Adorno, *Aesthetic Theory*, 14.

[10] *Aesthetic Theory*, 33.

is the most traumatic manifestation. In this context, elegy is one form in which art remains true to the idea of reconciliation, precisely by refusing total absorption in it. By this reckoning, radical art is almost by necessity epitaphic. It avoids the sacrilege of "cheerful art" in its writing of the wound that figures the absence of reconciliation, representing modernity in terms of memory and longing. This is at least one point at which Romanticism and critical theory might be said to intersect: their common commitment to a kind of non-identity that remains other to instrumental appropriation and that is both produced and threatened by the relentless movement of modernity towards a point at which difference is finally sublated.

Adorno refers the notion of the wound to Romanticism's appropriation of natural beauty in cultural landscape.[11] We get a very tangible sense of how cultural landscape is manifest in literature by looking at those moments in Romantic poetry at which elegiac and locodescriptive idioms merge. Geoffrey Hartman refers to these moments as indicative of a genre he names "nature-inscription."[12] In nature-inscription, mourning refuses to be confined to specific acts, sites or ritual occasions. As Hartman writes, "mourning and memory converge as an infinite task" just as "the poet reads landscape as if it were a monument or grave."[13] Wordsworth's "Lucy Gray; or, Solitude" is an obvious and accessible example of this monumentalization of loss in landscape – a process in which landscape, or the writing of landscape, is rendered non-identical with itself, such that it becomes an allegory of the beyond. In the poem an absence – "Those footmarks, one by one, / Into the middle of the plank; / And further there were none!" – stands as a trace that alludes to a kind of plenitude that lies beyond the cognitive scope of the poem. The suppositional nature of the concluding stanzas, the deferral to local lore – "– Yet some maintain that to this day / She is a living child"[14] – has the same effect as the periphrasis that elsewhere in Wordsworth creates the sense that "under the words are ghostlier words, half-received figures or fragments that seem to be at once a part of the lost object yet more living than what is present."[15] Adorno is getting at something very similar when he writes "The telos of artworks is a language whose words cannot be located

---

11 Adorno, *Aesthetic Theory*, 61–64.

12 Geoffrey H. Hartman, *The Unremarkable Wordsworth* (London: Methuen, 1987): 31–39.

13 Hartman, *The Unremarkable Wordsworth*, 30, 40.

14 Wordsworth, "Lucy Gray; or, Solitude" (1800), *Poems*, ed. Hayden, vol. 1: 394.

15 Hartman, *The Unremarkable Wordsworth*, 29.

on the spectrum: a language whose words are not imprisoned by a prestabilized universality."[16] Nature-inscription, in this respect, represents the possibility of nature beyond our appropriation of it. In Wordsworth's work this is quite literally a matter of the beyond – it is emblematized in the graves of dead children who figure the dissolution of the human into the natural.

This attempt to write landscape as a site of mourning, such that the aesthetic appropriation of nature stands as a monument to a kind of reconciliation with nature that is properly absent in the artefact, occurs constantly throughout Romanticism. If the poetic production of cultural landscape or nature-inscription actually seeks to convert unmediated nature itself into the raw material of aesthetic synthesis, which it offers as an object of aesthetic contemplation, it also renders non-identity in the spectral dissonance that points to something it is not able to secure except as a trace or a tremor of memory. The possibility of reconciliation here does not reside in a straightforward pastoralism, but in what is figured in the landscape. The grave – a wound on the face of the land – emblematizes the absence of unity between the natural and the human in the purely artefactual nature-inscription (which in another sense speaks of culture's triumph over the natural). It is as if cultural landscape places an internal limit on the ability of aesthetic synthesis to incorporate natural beauty into regimes of objectification and consumption.

In critical discussions of the Romanticism of Australian poets like Charles Harpur and Henry Kendall, some sort of notion of failure is usually evident, as if the attempt to appropriate and replicate Wordsworthian conventions is doomed when confronted with the alien, more emphatically undomesticated landscapes of nineteenth-century Australia. Phrases like "Romantic disinheritance" and "the absence of Romanticism as negativity"[17] imply something like the Antipodean fall of Romanticism – the inappropriateness or failure of the conventions accompanying European nature-inscription and cultural landscape in the new world. This is compatible with accounts of the demonizing of unassimilable "natural" elements in a settler society driven by the paranoid aggression of colonization. While I do not discount the obvious complicity of Australian representations of landscape with colonization, I do want to come at the issue of this complicity without straightaway

---

[16] Adorno, *Aesthetic Theory*, 83.

[17] On "Romantic disinheritance," see Andrew Taylor, *Reading Australian Poetry* (St. Lucia: U of Queensland P, 1987): 22–35. The second phrase summarizes the thesis of Paul Kane's *Australian Poetry: Romanticism and Negativity* (Cambridge: Cambridge UP, 1996).

collapsing the aesthetic into the instrumental (the standard move of Anglo-American ideology critique). The work of Harpur and Kendall, and for that matter of Marcus Clarke, who might also be described, at moments, as an Australian Romantic, suggests a more complex process in which the conventions of nature-inscription are redeployed in a conflicted manner that starts to make evident the extent to which these conventions are so easily overdetermined by the forces of both commodification and colonization. These forces, moreover, seem to operate in a synchronous manner such that the logic of the commodity engenders representational paradigms that conveniently exploit the potentially abject nature of nature as non-identity, and thus render themselves compatible with narratives of colonization as domestication. Ultimately my argument is this, though it is an argument that goes beyond the possibilities of this piece: if we are to disentangle the issue of the environment or nature as an ethical concern from its imbrication with colonialism, we need to resecure a provisional sense of nature as a thing-in-itself from the dominant sense of nature as an object of instrumental appropriation. This is, of course, not a matter of Romantic revivalism, but of much more varied resistance to and critique of objectification in which we understand how a semiotics of nature became implicated in instrumental rationality. It is, nevertheless, Romanticism's fleeting refusal of the logic of objectification, or the sense that it was, for a moment, one step ahead of objectification, that allows it to function as the locus for some sort of ethical nostalgia in which we can think the environment outside of instrumental imperatives (which is also why the green movement can be dismissed as Romantic). Romanticism, of course, quickly ossified into a style that entered into a range of complicities evident in a poem like *The Excursion.* In a way, the nineteenth-century redeployment of nature-inscription in Australia, at what was literally the cutting edge of empire, is also the story of the erosion of the ethicality that coheres around Romanticism's elegiac idiom. The instrumental appropriation of nature is mirrored in the mutations that occur at the level of cultural production.

When we read Harpur and Kendall, for example, we find moments at which an elegiac impulse enables poems that register the violence of colonization and the devastation inflicted on indigenous society. Harpur's "An Aboriginal Mother's Lament" and Kendall's so-called Aboriginal death songs ("Urara," "Ulmarra" and "The Last of His Tribe," for example) write indigeneity as a space and experience defined by lament, although in the case of Kendall's poems there is no sense of direct white culpability. While these texts read as contrived in their use of poetic persona, both poets manage the conventions of nature-inscription much more exactly in their writing of settler alienation and deprivation in the face of a landscape that monumentalizes not just the (deferred) possibility of man reconciled with nature, but also the perhaps more emphatic possibility of the settler's defeat at the hands of a nature that is now irrevocably associated with the threat of indigeneity. Harpur's "The Creek of the Four Graves" is a good example of colonial nature-inscription overdetermined in this way. While the poem clearly attempts to replicate the language and the style of poems like "The Ruined Cottage" and "Tintern Abbey," and while, as the title indicates, its cue is the typically Wordsworthian inscription of loss in landscape, its narrative recounts the massacre of four settlers by an Aboriginal tribe in a way that plays upon settler fears about nature unassimilated into a particular economy of value and use. The poem opens with an unambiguous statement of economic stakes: the settlers in the poem are in search of "streams and wider pastures" for "augmenting flocks and herds."[18] While Wordsworthian descriptions of nature as haunted, as spectral, as uncannily animated, seem to sit uncomfortably with the subsequent explosion of the bush into a horde of "stript / And painted savages" (167), there is a sense in which the shift from nature as object of reverence, as object of disinterested aesthetic contemplation, to nature as a site of indigenous violence and settler paranoia is actually mediated by the conventions of nature-inscription themselves. Initial descriptions of the "echoes" of the bush, of "the strange words moulding a strange speech" (163), are quite literally echoed later in the poem when the narrator describes the yells of "Wild men whose wild speech had no word for *mercy*!" (169). These lexical correspondences really organize the poem: the "dark arms of the forest trees," "the circling forest trees [...] carved from a crowded mass" (164) prefigure the "black shadows of untimely death" (168) with which the poem culminates, while the typically Wordsworthian use of prosopopoeia to represent the landscape as animated with the ominous presence of something otherworldly is rendered almost parodically literal when the settlers are attacked.

But to read this poem just as a document of settler paranoia, as a record of Romantic conventions in the service of colonial ideology, is to ignore the less obvious but no less operative dynamic by which these conventions are

---

[18] Charles Harpur, *The Poetical Works of Charles Harpur*, ed. Elizabeth Perkins (Melbourne: Angus & Robertson, 1984): 161. Further page references are in the main text.

themselves implicated in a dialectical relationship with the text as commodity. In "The Creek of the Four Graves," Romantic conventions are shown to generate their commodified doubles through a kind of hypostatization that seems actually internal to or immainent within them. Notions of haunting, of spectrality, of nature as a vast graveyard indicating a world beyond our cognitive scope, are, in other words, pushed to their literal extreme in the idea of the Aboriginal attack as an infernal eruption – "Hell's worst fiends burst howling up / Into the death-doomed world" (166). As Ian McLean suggests of this kind of writing, there is a very obvious gothicization at work here, as Romantic nature-inscription moves very easily into the more popularized forms of a *Blackwood's* tale of terror, a Gothic romance or a colonial adventure novel. In this gothicization the spectrality evident in elegy gives way to the prefabricated and prosaically packaged terror of literary sensationalism:

> The predilection for the ghostly presence of primitive forces found, for example, in the writing of Harpur, Kendall and Clarke, is typical of the interest in the occult which accompanied the symbolism and adventurism of late nineteenth-century art and literature, and provided a potent arena of imperial novels well into the twentieth century.[19]

McLean's reading of what he interestingly calls a symbolist aesthetic focuses on the ways in which the gothicization of the landscape provided nineteenth-century writers and artists with a way of "taking possession of the country," partly by writing indigeneity as spectral. What McLean does not discuss, and what I think is indexed in his naming of this aesthetic as symbolist, is the dynamic process in which the colonial appropriation of landscape also coincides with the commodification of Romantic conventions, which culminates in the formulaic production of tropes designed to capture the *frisson* of spectral alterity. In this process, the symbol itself becomes the objectified emblem of non-identity anticipating not reverence for a lost object, but horror at the thought of its infernal re-animation.

In Kendall's poems this dynamic is clearer, because they more obviously and unashamedly employ easily recognizable and derivative conventions that mark the shift from nature-inscription to Gothic phantasmagoria. In a number of Kendall's poems, Gothic evocations of fearfully embodied landscape use rhyme schemes, metre and refrains that are almost identical to those that

---

[19] Ian McLean, *White Aborigines: Identity Politics in Australian Art* (Cambridge: Cambridge UP, 1998): 50.

we find in symbolist landmarks like Poe's "The Raven" and "Annabel Lee." In these poems an ethical reverence for nature, still evident in Harpur, has given way to a hyperbolic and much more mannered gothicization in which the final effect of horror is clearly a function of Kendall's commitment to a set of conventions that were increasingly the signs of one's fraught integration into a competitive literary marketplace – a marketplace in which a certain kind of saleable thrill cohered in the wilful perversity that we find in a poem like "Cooranbean." In this poem, a glen is turned into a gaping abyss that embodies the abject, the obscene, the corporeal in a way that threatens some equally staged sense of patriarchal authority:

> A furlong of fetid back fen, with gelid green patches of pond,
> Lies dumb by the horns of the Glen – at the gates of the Horror beyond;
> And those who have looked on it tell of the terrible growths that are there –
> The flowerage fostered by Hell – the blossoms that startle and scare;
> If ever a wandering bird should light on Gehennas like this,
> Be sure that a cry will be heard and the sound of the flat adder's hiss.
> But, hard by the jaws of the bend is a ghastly Thing matted with moss –
> *Ah, Lord! be a father, a friend, for the sake of the Christ of the Cross!* [20]

This last line evokes the refrain in Coleridge's "Christabel," in which the poet urges God to protect Christabel from the morbid sexuality and witchcraft of Geraldine. In its possible feminization of the glen, the poem carries a trace memory of Coleridge that suggests its indebtedness to metropolitan culture and conventions, as well as its orientation to regimes of cultural capital and literary fashion. The poem is, in this sense, similar to Marcus Clarke's evocation of the "weird melancholy" of the Australian bush,[21] in which descriptions of the grotesque and the ghostly reveal, in their excessiveness, the pleasure of commodification. Clarke's accounts of "weird melancholy" are very obviously pastiches of Romantic, Gothic and symbolist motifs, such that nature is virtually a pretext around which the pleasure of citation is activated. A story like "Pretty Dick" works like this – the reader unearths a series of canonical signatures, from Wordsworth to Poe, embedded in its phobic representation of landscape. What I think we are reading in Kendall's poem is the point at which the poetic rendering of land-

---

[20] Henry Kendall, *The Poetical Works of Henry Kendall*, ed. T.T. Reed (Adelaide: Libraries Board of South Australia, 1966): 168.

[21] Marcus Clarke, *For the Term of His Natural Life, Short Stories, Critical Essays and Journalism*, ed. Michael Wilding (St. Lucia, Brisbane: U of Queensland P, 1976): 645.

scape is now solely at the mercy of the mutually sustaining forces of commodification and colonization. The *frisson nouveau* of the Gothic – already readable as a dead tissue of poetic effects – segues into a pastoralist attitude to nature in which natural beauty becomes a matter of picturesque vistas compatible with the imperatives of instrumental appropriation, while nature that remains unassimilated is abject in its ugliness, though no less emphatically commodified. In this shift, a possible environmentalism latent in nature-inscription vanishes into a notion of nature that is dichotomized into conventional notions of beauty and ugliness and wholly at the mercy of the culture industry. It is here that the object of critique implied by a cultural politics that is both environmental and anti-colonial comes into view: a semiotics of nature overwritten by the mutually sustaining forces of colonization and commodification.

WORKS CITED

Adorno, Theodor W. *Aesthetic Theory*, tr. Robert Hullot–Kentor (Minneapolis: U of Minnesota P, 1997).

Bennett, Tony. *Culture: a Reformer's Science* (Sydney: Allen & Unwin, 1998).

Clarke, Marcus. *For the Term of His Natural Life, Short Stories, Critical Essays and Journalism*, ed. Michael Wilding (St. Lucia, Brisbane: U of Queensland P, 1976).

Dobson, Andrew. "Critical Theory and Green Politics," in *The Politics of Nature: Exploration in Green Political Theory*, ed. Andrew Dobson & Paul Lucardie (London: Routledge, 1993): 190–209.

Harpur, Charles. *The Poetical Works of Charles Harpur*, ed. Elizabeth Perkins (Melbourne: Angus & Robertson, 1984).

Hartman, Geoffrey H. *The Unremarkable Wordsworth* (London: Methuen, 1987).

Hunter, Ian. "Aesthetics and Cultural Studies," *Cultural Studies*, ed Lawrence Grossberg, Cary Nelson & Paula Treichler (London: Routledge, 1992): 347–72.

Kane, Paul. *Australian Poetry: Romanticism and Negativity* (Cambridge: Cambridge UP, 1996).

Kendall, Henry. *The Poetical Works of Henry Kendall*, ed. T.T. Reed (Adelaide: Libraries Board of South Australia, 1966).

McLean, Ian. *White Aborigines: Identity Politics in Australian Art* (Cambridge: Cambridge UP, 1998).

Schama, Simon. *Landscape and Memory* (London: Fontana, 1995).

Soper, Kate. *What is Nature? Culture, Politics and the Non-Human* (Oxford: Blackwell, 1995).

Taylor, Andrew. *Reading Australian Poetry* (St. Lucia & Brisbane: U of Queensland P, 1987): 22–35.

Wellmer, Albrecht. *The Persistence of Modernity: Essays on Aesthetics, Ethics and the Postmodern*, tr. David Midgley (Cambridge MA: MIT Press, 1991).

Wordsworth, William. *The Poems*, ed. John O. Hayden, 2 vols. (Penguin English Poets; Harmondsworth: Penguin, 1977).

"Transported Landscapes"
Reflections on Empire and Environment in the Pacific[1]

RUTH BLAIR

> We cannot restore a past society, even if the haze of history hides its evils from us; we must rebuild society for ourselves, learning from the past what lessons and what warnings we are capable of learning.[2]

IN 1829, FIFTY-ONE YEARS after the first European landfall in the Hawai'ian islands, the Reverend Charles Stewart saw growing in the missionary compound in Honolulu: breadfruit; cape gooseberry; coffee plants; grapevines; tamarind; fig; lime; lemon; orange; mango; strawberries and the "pride of Barbadoes" (*Caesapinia coriaria*).[3] The first-mentioned (*Artocarpus altilis*) is the only indigenous plant on the list. The last two had grown from seeds Stewart himself had brought on a previous

---

[1] Thanks to Humphrey McQueen, Greg Dening and colleagues in the English Department of the University of Queensland for helpful comments on this essay. The phrase "transported landscapes" is from A. Grenfell Price, *The Western Invasions of the Pacific and its Continents* (Oxford: Clarendon, 1963): 188.

[2] R.M. MacIver, foreword to Karl Polanyi, *The Great Transformation* (1944; Boston MA: Beacon, 1957): x.

[3] In this discussion I give common European names for plants and, where I feel it may be useful, the European botanical names. I would have liked to have included a section on naming but space did not permit. There is a chapter on local plant names and a section on botanical names in Elmer D. Merrill's *Plant Life of the Pacific World* (New York: Macmillan, 1945).

trip (pride of Barbadoes) or sent from home (strawberries).[4] There is a pride here, as elsewhere in the Pacific, on the part of the colonists in the naturalization of plants from "home" and, more generally, in the successful transplantation of species. Stewart's delight echoes the confidence of Thomas Jefferson's opinion that "the greatest service which can be rendered any country is to add a useful plant to its culture."[5] The cultural as well as the physical "naturalization" of plants was a rapid process. Another fifty years on, Henry M. Whitney's *Hawaiian Guide Book* (1875) catalogues a riot of species in an irrevocably altered landscape:

> A few [trees] are indigenous, such as the cocoa-nut palm, the lauhala or screw palm (pandanus), the breadfruit, the ohia or native apple, the koa, the hau and the kukui or candle-nut tree; but many of the handsomer trees have been introduced from foreign countries, and have grown into magnificent stature, within the past quarter century. Among these are the mango, [...] the tamarind, the Chinese orange and the sweet orange, the lime, the alligator pear, the citron, the custard apple, the fig, the coffee, bananas, papaias, peaches, date-palms, magnolias, algarobas, and samang or monkey-pod. [...] Also several varieties of acacia, the eucalyptus of Australia, the brilliant poinciana regia, Norfolk and Caledonia pines, the royal and fan-palms, the Indian banyan, the bamboo, the loquot and Chinese plum, with the pepper, cinnamon and spice trees.[6]

The endemic or indigenous, in the frame of perception of the colonizers, was far from being valued for its own sake.[7] Rather, local species were valued either as potential commodities or for aesthetic qualities that fitted into the picture of the South Seas paradise that early explorers had constructed. The breadfruit and the coconut became emblems of both the bounty and the beauty of this paradise, widely illustrated and described. Herman Melville, embroidering an account by the missionary William Ellis,

---

[4] Charles S. Stewart, *A Visit to the South Seas in the United States Ship "Vincennes", During the Years 1829 & 1830* (1831; New York: Frederick J. Praeger, 1970): 66..

[5] Alfred W. Crosby, *Germs, Seeds and Animals: Studies in Ecological History* (Armonk NY: M.E. Sharpe, 1994): 18.

[6] Henry M. Whitney, *The Hawaiian Guide Book for Travellers* (Honolulu: Henry M. Whitney, 1875): 14–15

[7] "An *endemic* species is one that is native of, but confined to, a definite geographic area, such as a single island or part of an island, while an *indigenous* species is one that has attained [a] wider geographic range [...] The latter is, of course, a 'native' species wherever it occurs but may be of a very wide geographic distribution"; Merrill, *Plant Life of the Pacific World*, 63.

devotes a chapter to the coconut (*Cocos nucifera*) in *Omoo* (1847). He describes the "high estimation" in which the fruit is held:

> The cocoa-palm is [...] a tree by far the most important production of Nature in the Tropics. [...] Its very aspect is imposing. Asserting its supremacy by an erect and lofty bearing, it may be said to compare with other trees, as a man with inferior creatures.[8]

Ellis, who gave the first extensive account of Polynesian geography and customs, is generally more concerned with utility than with beauty, but he does spare a lyrical moment for the breadfruit tree to describe the effect of this "remarkably beautiful" tree in the Tahitian landscape:

> There is nothing very pleasing in the blossom; but a stately tree, clothed with dark shining leaves, and loaded with many hundreds of large light green or yellowish-coloured fruit, is one of the most splendid and beautiful objects to be met with among the rich and diversified scenery of a Tahitian landscape.[9]

The breadfruit may not have become as well known in the West as the coconut because the fruit never became as widely accepted in the Western diet. But Bligh's attempt to establish the tree in the West Indies and the famous mutiny on the *Bounty*, en route to the Caribbean with the seedlings, have assured its emblematic status. Bligh is remembered, on his tombstone, as an Admiral and a navigator and, wrongly, as the man "who first transplanted the bread fruit tree from Otaheiti to the West Indies." He failed to do that; the first act of the mutineers was to toss all the seedlings overboard.[10]

Sandalwood had a different fate. An aromatic but not a beautiful plant by European standards,[11] it was not so clearly identified with the Pacific, being found also in parts of Asia. Voracious European and American

---

8 Herman Melville, *Omoo* (1847; New York: Hendricks House, 1969): 262. On Melville's extensive use of Ellis in his three South Seas "romances," see Charles R. Anderson, *Melville in the South Seas* (New York: Columbia UP, 1939).

9 William Ellis, *Polynesian Researches* (New York: J.J. Harper, 1833), vol. 1: 45.

10 Greg Dening, *Mr Bligh's Bad Language: Passion, Power and Theatre on the Bounty* (Cambridge: Cambridge UP, 1992): 88.

11 "The sandalwood tree is not imposing: from about six or eight feet of fairly slender grey trunk, branches begin to straggle upwards at irregular intervals, bearing small shiny leaves, oval or pointed, depending on the species"; Dorothy Shineberg, *They Came for Sandalwood: A Study of the Sandalwood Trade in the South-West Pacific 1830–1865* (Melbourne: Melbourne UP & London: Cambridge UP, 1967): 1–2.

exploitation led to supplies being exhausted in Fiji by 1816. Marquesan sandalwood supplies were stripped in only 7 years (1814–21).[12]

Plants, of course, had moved themselves about the Pacific on the ocean and with the help of birds, and Pacific peoples had moved about, taking plants with them. The principal edible plants of Polynesia – the breadfruit, the coconut, the banana and the sweet potato – were all introduced.[13] Most are considered to have been brought from Indonesia and New Guinea by early Polynesian settlers. No one yet knows with any certainty how the sweet potato (*Ipomoea batatas*), of South American origin, found its way to the islands of the Pacific in pre-Columbian times. Sir Peter Buck (Te Rangi Hiroa) speculates that it was brought back by Polynesians voyaging to the coast of Peru.[14] And it was Polynesians, not Bligh, who first transported breadfruit seedlings. E.S.C. Handy included in his collection of Marquesan traditions one concerning "an expedition which set out for Rarotonga with

---

[12] Shineberg, *They Came for Sandalwood*, 7.

[13] The oldest recorded history of the coconut is in India. It is not clear whether it first found its way into the Pacific with man or by floating on ocean currents. "In the Pacific, it is scarcely possible to be sure that any island was unvisited by prehistoric voyagers, but some apparently virgin islands had thriving coconut groves"; Jonathan D. Sauer, "Reevaluation of the Coconut," in *Man Across the Sea*, ed. Carroll L. Riley et al. (Austin: U of Texas P, 1971): 313. Nevertheless, pre-Columbian planting and use of the species are established for the western and central Pacific islands; Jonathan D. Sauer, "Reevaluation of the Coconut," in *Man Across the Sea: Problems in Pre-Columbian Contacts*, ed. Carroll L. Riley et al. (Austin: U of Texas P, 1971): 317.

[14] Peter H. Buck (Te Rangi Hiroa), *Vikings of the Sunrise* (New York: Frederick A. Stokes, 1938): 307, 313–15. "The assessment [...] of the Pacific distribution which gave the Polynesian plant prehistoric status is unarguable"; D.E. Yen, "Distribution of the Sweet Potato," in *Man Across the Sea: Problems in Pre-Columbian Contacts*, ed. Carroll L. Riley et al. (Austin: U of Texas P, 1971): 335. Radiocarbon dating of pollen verifies the presence of the "aboriginally introduced" sweet potato (*Ipomoea batatas*) in the southern Cook Islands circa A.D. 640; Annette Parkes, "Environmental Change and the Impact of Polynesian Colonization: Sedimentary Records from Central Polynesia," in *Historical Ecology in the Pacific Islands: Prehistoric Environmental and Landscape Change*, ed. Patrick V. Kirch & Terry L. Hunt (New Haven CT: Yale UP, 1997): 178. There is evidence of the sweet potato "in Polynesia in prehistory and in Melanesia only in post-Columbian times"; D.E. Yen, "The Origins of Subsistence Agriculture in Oceania and the Potentials for Future Tropical Food Crops," *Economic Botany* 47.1 (1993): 3.

a ship loaded with young breadfruit plants."[15] The result was a considerable consistency of plant life across the islands, particularly of edible plants.

The Europeans and, in the case of Hawai'i, the Americans, did the moving about of plants – just as they cut the sandalwood – at a much faster rate than had previously occurred, dramatically altering agricultural practices. In his call for Western society to think about implications of imperial expansion for land management, A. Grenfell Price compared the "invasions of plants and animals" to the "sweeping and ruthless nature of the influx of those other 'exotics' – Western peoples with their diseases and culture."[16] Responding, in 1963, to the developing idea of the ecosystem, he saw that "the ravages committed by man subvert the relations and destroy the balance which nature has established between her organized and inorganic creations."[17] Though such a comment betrays a tendency to see pre-European-contact Polynesia as a wild ("natural") place, it nevertheless points to the fact that the most dramatic changes to the land have taken place as a result of European contact. Wherever human beings have gone, they have had an impact on ecosystems. Islands are particularly sensitive ecosystems and pre-European-contact species extinctions in the islands of the Pacific are well documented.[18] The ecosystems of the islands were modified in other ways too. In New Guinea, for example, agricultural practices before European contact produced extensive grasslands and the entry of the sweet potato also produced changes in the landscape.[19] Nevertheless, the Pacific islands are of interest to those with environmental concerns today because the changes to the various island ecosystems post-European contact were accelerated and dramatic.

Today we are aware of the danger of extinction, not of the breadfruit and the coconut, which were already well-established food crops pre-European contact, but of many other endemic species in the islands of the

---

15 Buck, *Vikings of the Sunrise*, 308.

16 A. Grenfell Price, *The Western Invasions of the Pacific and its Continents* (Oxford: Clarendon, 1963): 176.

17 Price, *The Western Invasions of the Pacific*, 177.

18 Patrick V. Kirch & Terry L. Hunt, *Historical Ecology in the Pacific Islands: Prehistoric Environmental and Landscape Change* (New Haven CT: Yale UP, 1997): 53 and passim; I.G. Simmons, *Changing the Face of the Earth: Culture, Environment, History* (Oxford: Blackwell, 1989): 123. See also *Man's Place in the Island Ecosystem*, ed. F.R. Fosberg (Honolulu: Bishop Museum Press, 1965).

19 Kirch & Hunt, *Historical Ecology in the Pacific Islands*, 12, 47–50.

Pacific.[20] Melville commented on the potential for the introduced guava bush to "cover the entire island" of Tahiti, but he said this without too much concern, in rather the cavalier tone with which his contemporaries lamented the inevitable extinction of the indigenous people of the area.[21] We believe, today, that we have a superior view and care about biodiversity, and in developing an understanding of and concern for biodiversity we are retrieving knowledge of earlier agricultural societies. But we should be wary of applauding our efforts and scapegoating the past. It is too easy to assume the rightness (and the righteousness) of the environmental view. What concerns me here is to identify how and why a knowledge of past practices can be useful to us today and to ponder what questions it is valid to ask about colonial practices from where I stand – an early-twenty-first century settler Australian concerned about environmental degradation. I realise that I am going over ground already well trodden by environmental historians here, but in relation to the Pacific I want to ask the questions again. If we do not understand exactly why we are interrogating the past here, we can fall into the (now) too easy position of lamentation, which was tactically useful as a wake-up call in 1963 but which today simply massages guilt rather than producing a new vision.

I attempt to gauge, here, the usefulness of investigating the past both for a global community and for indigenous inhabitants of the Pacific islands. I try to establish a means of evaluating two views of human exploitation of the resources of the non-human world: that of Western capitalism's "development" agenda; and the more conservation-oriented interest in "sustainability" which pays attention to a network of relationships of people and land. "Sustainability" is something of a cliché these days, but it is about the only term we have for a balanced management of the land. My focus, therefore, is not so much on how the heady transportation of species in the early days of settlement in the islands produced a loss of both species and cultural identity, but on a general failure, in those early days, to see a network of relationships with the land. I have no intention of repeating the gesture of appropriating indigenous practice as a critique of Western capitalism – a gesture that was part of the duplicitous colonialist

---

[20] For lists of extinct and endangered species, see Arthur L. Dahl, *Regional Ecosystems Survey of the South Pacific Area* (Noumea, New Caledonia: South Pacific Commission, 1980).

[21] Melville, *Omoo*, 261.

baggage. My contention is that what we (members of a Western capitalist society) can learn by interrogating the past is something about attitude and attention. The past in the Pacific can show us a lesson not already learned about alternative ways of doing things. It can show us in full, glorious colour the outcomes of utter disregard for other ways than those of Western capitalism. That we badly need this lesson is evident in the imperialistic rhetoric and operations of, first, the "green revolution" and, second, those today championing biotechnology.

Histories of the European and American invasion of the Pacific (even recent ones) say little about the land itself, apart from the scientific interest in new species and a strong interest in fitting the islands into an evolving Western aesthetic tradition.[22] Few historians speak in detail of cultivation practices or of the relationship between people and the land. This omission is not specific to Pacific history nor, indeed, to the discipline of history. Edward Hyams's remarkable *Soil and Civilization* (1952) was an attempt to redress the historical gap. Johnson D. Hill and Walter E. Stuermann take on philosophy: "One of the striking features of the history of philosophy and of Christian thought is the almost total absence of reflection on agriculture, agrarianism, and the significance of farm labor."[23] What we have to work with, then, in searching for the history of land use, are glimpses of what Tom Griffiths calls the "incidental and discounted dimension of imperialism."[24]

Anthropology, as we would expect, does not ignore agricultural practices.[25] Malinowski's extensive work on agriculture in the Trobriand

[22] The first great European intellectual history of the encounter is Bernard Smith's *European Vision and the South Pacific* (Oxford: Clarendon, 1960). The first great history of the Pacific islands to attempt to recreate the story from the other side of the beach is Greg Dening's *Islands and Beaches: Discourse on a Silent Land; Marquesas 1774–1880* (Chicago: Dorsey Press, 1980).

[23] Johnson D. Hill & Walter E. Stuermann, *Roots in the Soil: An Introduction to Philosophy of Agriculture* (New York: Philosophical Library, 1964): ix.

[24] *Ecology and Empire: Environmental History of Settler Societies*, ed. Tom Griffiths & Libby Robin (Melbourne: Melbourne UP, 1997): 2.

[25] See Clifford Geertz, *Agricultural Involution: the Processes of Ecological Change in Indonesia* (Berkeley: U of California P, 1963). Noteworthy, too, is Frederick Rose's study,

islands, published as *Coral Gardens and their Magic* (1935), positions us to see and to take seriously the way Trobriand islanders manage their land and the inextricable involvement of agriculture with other social practices. He begins, like Melville in his Marquesan narrative, *Typee* (1846), with the view from the shore: "Nothing is perhaps more impressive to an ethnographer on his first pilgrimage to the field than the overwhelming force of vegetable life and the apparent futility of man's efforts to control it." The cadences of the opening paragraph roll on towards their deeply nineteenth-century conclusion:

> Nature here seems not yet to have been subdued by man and fashioned to serve his purposes. Man, on the contrary, is but part of her scheme, sheltering precariously under what the jungle has yielded, clad in dried leaves, subsisting on that which, year after year, he wrests from the virgin forest and which, after a few years, returns to it again.[26]

But as we turn the page, we move from the impersonal third person to direct address, the narrative drawing us, Conrad-like, closer into the scene:

> If you were to settle in one of the hamlets and follow the work and interests of the natives, the perspective would change considerably. You would find everywhere that agriculture is a businesslike procedure, that it is not merely a highly skilled and technical enterprise but also an important ceremonial of the tribe; that the whole territory is well marked, legally defined and more or less appropriated to individuals or groups.[27]

Malinowski has provided in this book an invaluable record of Trobriand island agricultural practices and has exploded a set of assumptions about the wildness of a people and their land.

Taking agriculture seriously means crossing disciplinary boundaries and getting down in the dirt. Clifford Geertz homes in on the somewhat vexed relationship of his discipline with agriculture, showing how the work of "agricultural geographers" too easily resorted to a crude functionalism and produced only the "gross" questions of influence of environment on culture or vice versa. In *Agricultural Involution*, his 1963 study of the history of agriculture in the islands of Indonesia, framed, like Price's 1963 history, by

---

based on earlier research: *The Traditional Mode of Production on the Australian Aborigines* (Sydney: Angus & Robertson, 1987).

26 Bronislaw Malinowski, *Coral Gardens and their Magic*, 3 vols. (1935; London: Allen & Unwin, 1966): 3.

27 Malinowski, *Coral Gardens and their Magic*, 4.

the concept of the ecosystem, he "attempts to achieve more exact specifications of the relations between selected human activities, biological transactions, and physical processes."[28] Like Malinowski before him, he gives us details of the practices, describing *swidden* (more commonly "slash and burn", though this terminology has negative connotations that are, as Geertz shows, not always merited) and *sawah* (irrigated terrace) agriculture. He has "learned" agriculture and grounds his history in the documentation of practice.[29] Studies like Malinowski's and Geertz's offer immensely important "glimpses" for reflection on current practice. Of course, the recent area of environmental history whose roots go back to G.P. Marsh's *Man and Nature* (1864) crosses disciplinary boundaries and draws us to see the intricacies of the relationship between cultures and environments.[30]

As we look to the history of the Pacific, we have to work hard for our glimpses of indigenous agriculture. Early visitors paid little attention to agricultural practices and to the multifarious uses of endemic plants, preferring to bestow on the local environment the great "advantages" of introduced species. To them, the indigenous inhabitants were a lazy, "fallen" people who had no need to work because nature's bounty was all around them. "They have no other care," said La Pérouse (famously), "but that of [...] gathering without labour the fruits that hang over their heads."[31] Like the breadfruit and the coconut, such a way of life could be inscribed in the European imagination as paradisal. But missionaries led the way in planting in it the seeds of depravity. Either way, the landscape,

---

[28] Geertz, *Agricultural Involution*, 3.

[29] Michel de Certeau has helped me see the value of "ways of using things or words according to circumstances" – the detail Malinowski and Geertz offer us; Michel de Certeau, *Heterologies: Discourse on the Other*, tr. Brian Massumi (Minneapolis: U of Minnesota P, 1986): 20. Geertz's theory of "agricultural involution" is grounded in the dirt and the detail.

[30] More recent works in environmental history include Clarence Glacken's *Traces on the Rhodian Shore: Nature and Culture in Western Thought From Ancient Times to the End of the 18th Century* (Berkeley: U of California P, 1973), Roderick Nash's *Wilderness and the American Mind* (New Haven CT: Yale UP, 1973), Carolyn Merchant's *The Death of Nature: Women, Ecology and the Scientific Revolution* (San Francisco: Harper & Row, 1980) and *Ecological Revolutions: Nature, Gender and Science in New England* (Chapel Hill: U of North Carolina P, 1989), Alfred W. Crosby's *Ecological Imperialism* (1986), and I.G. Simmons's *Changing the Face of the Earth* (1989).

[31] Smith, *European Vision and the South Pacific*,103.

like the people, was conceived of as wild. Few early visitors saw methods by which the Polynesians themselves had worked the land and transformed the landscape. Johann Reinhold and Georg Forster, naturalists on Cook's second voyage, are exceptions. Patrick V. Kirch, in his introduction to *Historical Ecology in the Pacific Islands* (1997), notes how their observations run counter to the idea of an untamed paradise and a fallen people. They recognized that "the breadfruit (*Artocarpus altilis*), the Tahitian chestnut (*Inocarpus fagiferus*) and the vi apple (*Spondias dulcis*) trees of the Tahitian landscape were not part of an original Garden of Eden but had been brought to the islands by humans who planted them."[32] Johann Reinhold Forster's *Resolution* journal is also important for its information about indigenous agriculture. Forster describes irrigation methods for the growing of taro in Tahiti: "The natives had made several wears [*sic*] of stone across the river in order to raise & to stem the water & by that means introduce it into their plantations of *Taro* or *Arum esculentum* [*Colocasia esculenta*]."[33] In Tonga, he observes the carefully tended gardens:

> I went into a Garden or place, which was inclosed with reeds artfully plaited diagonally. [...] Within were Bananas & other trees [...] All these Inclosures connect & I could walk through more than 10 along the Shore.[34]

Further up the country:

> [We] passed through many plantations, all enclosed as before mentioned, or with a hedge of *Erythrina corallodendrum*; then we came into a lane between plantations, which had Plantains & Yams planted in regular rows, afterwards in a plain with Grass, & from thence into a walk of 4 rows of Coconut Trees of about a Mile long & then into a land between plantations of the greatest regularity & surrounded by Shaddocks [pomelo or *Citrus paradisi*] & other fine trees.[35]

Greg Dening gleans from the historical record glimpses of the management of breadfruit in the Marquesas that show the intertwining of agriculture and social practice:

---

[32] Patrick V. Kirch, introduction to *Historical Ecology in the Pacific Islands*, 5.

[33] J.R. Forster, *The Resolution Journal of Johann Reinhold Forster 1772–1775*, ed. Michael E. Hoare, 4 vols. (London: Hakluyt Society, 1982), vol. 2: 341.

[34] Forster, *The Resolution Journal*, vol. 3: 378.

[35] *The Resolution Journal*, vol. 3: 379.

> Breadfruit grew in Te Henua [the islanders' name for their island world] with extraordinary variety and vigour. It grew on the valley floors, up the slopes and deep in the valley reaches, fruiting two and three times a year. Enata [Marquesans] could name more than thirty varieties of breadfruit; they divided their year into its four seasons, set their mathematical systems by its bundles of cropped fruit, associated their astronomic observations with its growth and waning, sang songs of its beauty and strength, hedged around every step of its planting and cropping with supernatural protection, killed those whose misbehaviour threatened it, and destroyed their enemies by destroying their trees. [...] Enata did not cultivate it in any strict sense. They planted it and cropped it. They owned trees in isolated valleys, on the slopes of hills, and around their houses, but they did not parcel up land to make orchards of them. They set a young tree in the ground for a new-born child. Their worst enemy was he who by curse or bad action could dry up the rains.[36]

Such discussions of indigenous agriculture are rare in twentieth-century histories. I must query, however, Dening's questioning of the status of indigenous management. Though less obviously organized than Forster's Tongan gardens, this thoughtful husbandry seems like cultivation to me. Who knows the reason for this "helter-skelter" (Geertz's term) planting? Annie Walter, writing on traditional tree farming in Vanuatu, shows how we are only just beginning to understand indigenous agriculture in the Pacific. At first glance, she says, "the management of fruit trees may appear disorganized", but it is grounded in "a wide understanding of nature and its biological rhythms." Based on diversity, the system "requires the management of a large number of plant types." The reason for the preservation of diversity is that, though frequent hurricanes may destroy a whole crop, "it is unusual for all the specimens of all the usable species to be destroyed at once." The fruit trees are protected through a knowledge that is passed from one generation to another.[37] We should also note, that far from giving in to bad seasons, Pacific islanders had developed a unique method of preservation by fermentation. W.G.L. Aalsberg and Susan Par-

---

[36] Dening, *Islands and Beaches*, 48. Dening's account is derived from the missionary/beachcomber, William Pascoe Crook, the beachcomber Edward Robarts, and various early visitors.

[37] Annie Walter, "Knowledge for Survival: Traditional Tree Farming in Vanuatu," in *Fauna, Flora, Food and Medicine*, vol. 3 of *Science of Pacific Island Peoples*, ed. John Morrison, Paul Geraghty & Linda Crowl (Suva, Fiji: Institute of Pacific Studies, 1994): 199.

kinson say it is likely that the procedure dates back 3,000 years.[38] Walter's and Aalsberg & Parkinson's articles appear in a recent multi-volume work on the science of Pacific Island peoples. Like Kirch and Hunt's study of pre-European contact ecology, this work, arising out of a 1992 conference on the subject, provides ways for us to begin to see and understand a tradition of indigenous management of the island ecosystems.

Those who settled or sojourned at length in the islands could hardly ignore some aspects of local cultivation practices. Ellis, who arrived in the Tahitian group of islands in 1817, twenty years after the *Duff* carried the first London Missionary Society missionaries to the Pacific, catalogued some of the major indigenous exploitations of plants – the various uses of the coconut fruit and palm, for example. In Melville's *Typee*, we have accounts (largely derived from Ellis) of indigenous exploitation of resources – the gathering of food, the making of tapa (a cloth from the bark of the paper mulberry tree, *Morus papyrifera*). The Marquesas, to the north and east of Tahiti, were a special case. Difficult to colonise, they were still, in 1842, what Europeans would call "behind".[39] Hence, Melville was able, with some legitimacy, to describe what were, to Western eyes, "primitive" agricultural customs. But twenty years on from Ellis's arrival, Tahiti was a different matter. European methods of agriculture were already well established when Melville toured the islands. His *Omoo, or Reef Rovings in the Pacific* tells the following story.

On his rambles, Melville's narrator, Omoo, encounters European and American planters. A Yankee and a Cockney (former 'seafaring men") are growing the Tombez sweet potato (Melville noting its Peruvian origin) to sell to passing ships:

> The cleared tract which they occupied, comprised some thirty acres, level as a prairie, part of which was under cultivation; the whole being fenced in, by a stout palisade of trunks and boughs of trees, staked firmly in the ground. This

[38] W.G.L. Aalsberg & Susan Parkinson, "Traditional Pacific Food Technology," in *Fauna, Flora, Food and Medicine*, vol. 3 of *Science of Pacific Island Peoples*, ed. John Morrison, Paul Geraghty & Linda Crowl (Suva, Fiji: Institute of Pacific Studies, 1994): 5.

[39] On the difficulties of establishing missionaries in the Marquesas, see Dening, *Islands and Beaches*, ch. 3 and 5. On aggressive behaviour, including wanton slaughter, on the part of other would-be colonists (the Spaniards in 1595; the American captain David Porter in 1813), see Dening, *Islands and Beaches*, 9–11, 26–31.

> was necessary, as a defense against the wild cattle and hogs overrunning the island.[40]

Omoo and his companion earn their keep grubbing out vegetation to clear more land for the commercial crop. At another point, Melville creates a dramatic episode out of the presence of "wild" (introduced) cattle. Though he is supplementing his narrative here (as was his wont) with information gleaned from another source and about another island (Hawai'i), the cattle are part of the picture he gives us, intermittently throughout this text, of changing agricultural practices.[41] *Omoo* is a rather sad, down-at-heel comedy, portraying a rather sad, down-at-heel Tahiti following colonization. The high romance of the South Seas, which the Marquesas makes still possible in *Typee*, is no longer available in Tahiti.

From Forster's account of Tahiti, we know there were deliberate plantings:

> The foot of the Hills are cultivated. The breadfruit Tree is a lofty one with large, beautifully jagged leaves, & large greenish fruit: between them are plenty of Bananas & Plantains, & now & then a piece of ground occupied by the purple eatable *Arum* [...]; every where is the *Dioscorea alata* or sweet Potato, twisting round trees & Shrubs. Now & then you see a large Apple tree [*Spondias dulcis*] [...] & a red flower stand mixed between them.[42]

---

[40] Melville, *Omoo*, 202. D.E. Yen cites Melville's naming of the Tombez potato as evidence for the influence of a later introduced South American species of the sweet potato on earlier varieties, making it possibly "responsible for a somewhat greater variation in modern Polynesian sweet potatoes than was evident in pre-European times"; Yen, "Distribution of the Sweet Potato," in *Man Across the Sea: Problems in Pre-Columbian Contacts*, ed. Carroll L. Riley et al. (Austin: U of Texas P, 1971): 335–36. Though the wild cattle episode (see below) and his frequent recourse to others' texts makes Melville not the most reliable of archaeological witnesses, the potato-growing farmers, Shorty and Zeke, do seem to have existed, being mentioned by subsequent travellers; *Omoo*, Notes, 404. Melville, who loved the concrete – and sometimes the overdetermined – gives us both the colour (yellow) and the botanical name.

[41] "This curious chapter [54: "Some Account of the Wild Cattle in Polynesia"] was developed mostly out of hints culled from Wilkes's account of Hawaii. [...] There are few references to such bullocks in the Society Islands, and possibly the whole bullock-hunting episode is a bubble blown up from Wilkes; *Omoo*, Notes, 405).

[42] Forster, *The Resolution Journal*, vol. 2: 336–37. *Dioscorea alata*: Forster was here using the botanical name for the yam. Columbus applied a West African term for yams to the South American sweet potato and confusion of the two tubers persisted; Donald

Straight lines and fences were not necessarily new to the islands, as Forster's Tongan descriptions tell us, but the potato-growing episode in *Omoo* shows once again an accelerated alteration of the landscape and a disregard of former practices. In Melville's Tahiti, sweet potato vines do not twist around trees or create that combination of utility and beauty that struck Forster and that modern permaculture attempts to achieve.[43] Within fifty years of European settlement, nothing is to be seen of the indigenous population's way of managing the land but some straggly patches of taro and yams and a few coconut groves on the shoreline.

This is the outcome of a process started by William Ellis and other colonizers. Ellis gives a succinct account of the introduction of capitalistic enterprise and of a concomitant change in agricultural practice. Having moved from the island of Eimeo in the Georgia group of the islands that make up present day Tahiti, to the island of Huahine in the Society group to the north, Ellis sets about altering lives and landscape:

> To increase their [the islanders'] wants, or to make some of the comforts and decencies of society as desirable as the bare necessities of life, appeared to us the most probable method of furnishing incitements to permanent industry. It was therefore recommended to them to erect for themselves more comfortable dwellings, and cultivate a larger quantity of ground, to meet the exigencies of those seasons of scarcity which they often experienced during the intervals between the breadfruit crops. We also persuaded them to adopt [our clothing and] our social and domestic habits of life. This required [...] a variety of articles that could not be supplied on the islands and must be obtained through the medium of commerce with Port Jackson and England; *and they could only procure these articles in a degree equal to that in which they multiplied the productions of the soil, so as to be able to exchange them for the manufactured goods of civilized countries.*[44]

D. Brand, "The Sweet Potato: An Exercise in Methodology," in *Man Across the Sea: Problems in Pre-Columbian Contacts*, ed. Carroll L. Riley et al. (Austin: U of Texas P, 1971): 351–52.

[43] "Permaculture" is the term for a design system, developed by Bill Mollison and David Holmgren, "for sustainable environments providing food, energy, shelter, material and non-material needs, as well as the social and economic infrastructures that support them" (*Permaculture International Journal* 72 (September–November 1999).

[44] William Ellis, *Polynesian Researches* (New York: J.J. Harper, 1833), vol. 2: 209–10 (my emphasis).

As the islands have no "spontaneous" valuable commodities, in comparison with "the sandal-wood of the Sandwich Islands and the pine-timber of New Zealand [...] Whatever articles of export they could ever expect to furnish, must be the product of their own industry." Consequently, the existence of an indigenous variety of cotton in the islands leads to the "successful" introduction of several new varieties. The women take to cotton growing, being "the most active in this work":

> [They] laboured diligently and perseveringly, cutting down in the mountains wood for the fencing, employing their own servants to transport it to the shore, clearing away the brushwood, enclosing the ground, digging the soil, planting the seed, watching with constancy its growth, and carefully gathering the cotton.[45]

In a nutshell, we have the replacement of a "polycultural" system of agriculture ("raising a diverse assemblage of crops in functional interdependence"[46]) by a "monocultural" system, and the alteration of a landscape with no care for existing relationships between people and land. Ellis speculates on reasons for the women's diligence – do they want what money can buy more than the men, or do they feel "more peculiarly their obligations to Christianity"? He is not, however, concerned with the answer and we have only his version of things with no account of the social complexities that drive these decisions (Ellis 2, 210). What he never questions is that the invader's way is superior. There is not (yet) any listening to the other culture. So it is also with agricultural systems. There is an implication that the people did not know how to manage difficult times between breadfruit crops because they were both lazy and ignorant. Set them (well, the women, at least) to work and ply them with superior knowledge.

Biological and archaeological investigations today are finding out more about pre-colonial agriculture. The recent *Historical Ecology in the Pacific Islands* (1997), a collaborative effort between archaeologists and natural scientists, is an invaluable contribution to our understanding of the interactions between people and their environments in the islands before the European invasion. Patrick Kirch reminds us, in his Introduction, that "from their very first incursions these populations began to change island

[45] Ellis, *Polynesian Researches*, vol. 2: 210.

[46] David R. Harris, "Agricultural Systems, Ecosystems and the Origins of Agriculture," in *The Domestication and Exploitation of Plants and Animals*, ed. Peter J. Ucko & G.W. Dimbleby (London: Gerald Duckworth, 1969): 5.

ecosystems in diverse ways"; but the myth that "pre-European inhabitants [...] were simply actors upon a changeless stage" has proved persistent.[47] A stress on the impact of Europeans and a down-playing of "the extent of earlier indigenous impact"[48] means that with the best will in the world, environmentally conscious Western scientists and commentators are perpetuating an imperialist gesture and failing to see the work and the relationships with the land that already existed. It is too easy to focus on the preservation of indigenous or endemic species when perhaps, like Ellis, we are still not prepared to attend to the nature of *relationships* with the land. It is not the introduction of foreign species and the displacement of endemic ones that, in itself, constitutes the problem. People had been affecting fragile island ecosystems in the Pacific since the very first arrival of "large omnivorous vertebrates."[49] I do not mean here to argue *against* saving endemic species. There are very good arguments (many of them cultural) for conservation. But the European invasion, compared with earlier invasions, was fast and dramatic, buoyed by the confident assumptions of Western capitalism. The real value of looking at the history of the impact of European contact on the land may be not so much in lamenting loss as in facing the motivations for Western practices in order to be able to identify and debate them today, for they are still going on. Glimpses of the past in the Pacific can teach us something about attending to other ways of growing crops to satisfy human needs – about the value of listening and seeing as Ellis and others did not.

A long tradition of attention to the flora and fauna of the Pacific has been aided by work in that area fostered by the Bishop Museum, founded in Honolulu in 1889. From the early days of European contact, attention was paid to the potential use value of the plants of the Pacific. "Economic botany", as it came to be called, produced various lists of plants throughout the nineteenth century. "Usefulness", then and now, is an attractive concept.[50] Otto Degner's 1940 study of the Fijian islands prefigures the

---

47 Kirch & Hunt, *Historical Ecology in the Pacific Islands*, 2–3, 4.

48 *Historical Ecology in the Pacific Islands*, 10.

49 *Historical Ecology in the Pacific Islands*, 2.

50 Catalogues like I.H. Burkill's *A Dictionary of Economic Products of the Malay Peninsula* (London: Publication on behalf of the Governments of the Straits Settlements and Federated Malay States by the Crown Agents for the Colonies, 1935) are remarkable for the breadth of resources on which they draw, both scientific and cultural. John R. Jackson's *Commercial Botany of the Nineteenth Century* (London: Cassell, 1890) is a shorter work

interest in the medicinal use value of many indigenous species. But the path to usefulness too easily invites the imperialist gesture. Here is modern ecological science in the voice of William Ellis:

> It is not possible to put a monetary value on each unusual or endemic species. [...] We do not know in advance what use, if any, might be found for each of these species. However there are many examples of rare or endemic species which have had very great economic importance. The Monterey pine (*Pinus radiata*) is an endemic species of the California coast with little local economic value; however, it has been introduced to New Zealand and other countries where it is now the basis for much of the forest industry. An obscure insect in one country may be found to be the ideal biological control for an important agricultural pest somewhere else. Allowing our natural environments, habitats, and species to be destroyed is rather like throwing away a box of rocks because we cannot tell which ones are worthless and which ones are uncut jewels.

This is from Arthur L. Dahl's Introduction to his 1980 compilation of South Pacific ecosystems,[51] endangered species and conservation efforts and planning. Contemporary environmentalism is speaking here (though let's hope we have learned a bit since 1980), but the *raison d'être* for the compilation differs little from Ellis's in his account of Polynesian geography and customs, and there is equal enthusiasm for introducing "useful" plants into other landscapes. (I am sure the mention of *Pinus radiata* will make a few contemporary hackles rise.) And it gets worse. Dahl goes on to champion conservation as an encouragement to tourism. Conservation, he says, means new jobs. It "also contributes to the quality of life of the local inhabitants by providing them with areas for rest and recreation where they can go to learn about the environment in which their traditional culture and island way of life evolved" (3). Ellis hardly gets any more paternalistic – or more circular.

Attention to a different approach to the management of the land and its plant species is important for the West – bearing in mind the dangers of repeating the colonial gesture. It is even more important for the islands themselves. Jack Golson ends a discussion of early agricultural practices in Papua New Guinea by drawing attention to the changes that are still making an impact on a threatened way of life:

---

that offers an introduction to the registering of plants according to their potential for economic exploitation.

[51] Dahl, *Regional Ecosystems Survey of the South Pacific Area*, 3.

> Today, in the Papua New Guinea Highlands the cash economy and rising populations are making new demands on natural and social systems and are requiring adjustments from them of a scale and urgency unparalleled in the past.[52]

With this in mind, we might consider studies of the Enga province of the Papua New Guinea Highlands, from Waddell's study of sweet-potato mounds to a recent thesis on land tenure in the area by Andrew Lakau, which show the value of studying an area that has not been as dramatically affected by agricultural transitions as other areas.[53] Enga has a lot to teach about those "balanced" relations with the land that Price sought.[54] Lakau looks at the ecological implications of traditional land tenure in the area, seeing (after Waddell) the tendency for people to live in "dispersed or scattered homesteads" (as opposed to villages) as "an adaptive strategy in relation to ecological limits."[55] "Through long experience," he says, Kaina farmers adopted "various ecological and cultural strategies to adjust themselves to the land and harsh physical environment" and have "developed a folk wisdom in choosing various garden types and cultivation techniques."[56] In other, more hospitable parts of the Highlands, good garden-

---

52 Jack Golson, "From Horticulture to Agriculture in New Guinea," in *Historical Ecology in the Pacific Islands: Prehistoric Environmental and Landscape Change*, ed. Patrick V. Kirch & Terry L. Hunt (New Haven CT: Yale UP, 1997): 50.

53 Though in many ways New Guinea, as part of Melanesia, is different from the other islands (Polynesia) I have mostly been discussing, I am using it as my example here because of shared plant species and because so much useful work has been done on areas where traditional agricultural practices have survived. This is not the case for many of the islands of the Pacific, but there are lessons to learn here about the conservation, not only of species, but of practices.

54 Price, *The Western Invasions of the Pacific*, 177. See also a study of the Chimbu people of the eastern-central Highlands by H.C. Brookfield and Paula Brown, *Struggle for Land: Agriculture and Group Territories Among the Chimbu of the New Guinea Highlands* (Melbourne: Oxford UP, 1963). There is a detailed description of sweet potato cultivation (46–47) and a confirmation of the value of studying traditional practices: "Considered in this way New Guinea highland agriculture is both technologically efficient and relatively successful. Although its long-term effects remain debatable, highland agriculture is nowhere destructive of resources in the short term, and the various techniques are adapted to local and environmental conditions" (167).

55 Andrew A. Lakau, "Customary Land Tenure and Alienation of Customary Land Rights Among the Kaina, Enga Province, Papua New Guinea" (unpublished doctoral dissertation, University of Queensland, 1994): 135.

56 Lakau, "Customary Land Tenure," 157, 138.

ing land has been taken up by "coffee plantings, cattle and the alienation of land by the State", motivated by the development of "a dynamic agricultural and trading capitalism" (147). In contrast, the Kaina have developed a "diversified and multiple cropping pattern". Soil preparation, weeding and harvesting are "done with reasonable care and in complete harmony with soil types, slopes and plant cover" (biochemical artificial fertilization and manuring are conspicuously absent – "the 'green revolution' is a long way off," leading to "a relatively high degree of subsistence security."[57] The Kaina "obviously apply a safety or survival first strategy before profit maximisation in resource use"; the "flexibility in residential pattern and of land use" which has "served to keep environmental balance" in the interaction of the Kaina with their land is (and this is the point of Lakau's thesis) an important lesson.[58] At a time when "ownership, use and management of Papua New Guinea's customary land tenure [is] coming increasingly under public scrutiny", account must be taken of the "complex and customary land tenure system" with its "multiplicity of rights and obligations."[59]

"Indigenous systems are cultural in their working, cultural in their results," says D.E. Yen in an article on subsistence agriculture in the Pacific and the potential therein for future food crops. This means, as Yen glosses it, "not conforming with the precepts of Ricardo, Keynes, or supply-side economics."[60] But subsistence systems of farming, such as those practised by the Kaina, are, as Marshall Sahlins points out, economic systems. It is in their difference from capitalist systems that they offer important lessons about both economics and agriculture.[61] They can teach us the value of seeing economics as something besides market forces.[62] A large part of the

---

57 Lakau, "Customary Land Tenure," 146, 147. The "green revolution" is the paradoxical term for Western capitalism's influence on indigenous agriculture in the post-war period.

58 "Customary Land Tenure," 146.

59 "Customary Land Tenure," 157, 362a.

60 Yen, "The Origins of Subsistence Agriculture in Oceania," 4.

61 "The Origins of Subsistence Agriculture in Oceania," 4; Marshall Sahlins, *Stone Age Economics* (Chicago: Aldine Atherton, 1972): 1–39.

62 "The outstanding discovery of recent historical and anthropological research is that man's economy, as a rule, is submerged in his social relations. He does not act so as to safeguard his individual interest in the possession of material goods; he acts so as

world lives within the capitalist hegemony and we ignore capitalist economics at our peril. But environmental concerns and that relatively new concept of 'sustainability' are leading us to consider other relationships (the cultural ones; meeting the needs of a continuing human relationship with the land) besides those of market-to-market. Ellis gave no thought to the potential value of the intertwined cultural and agricultural practices of 'his' islanders. If the history of agriculture in the Pacific matters today to a wider world, it is because the new imperialists (the biotechnological "developers") are also not listening.

D.E. Yen asks those concerned with economic botany to consider not new, genetically modified crops, but indigenous plants, part of an earlier subsistence agriculture, as having potential as important food plants of the future in tropical areas. Instead of the introduced and now global 'staple' plants, he asks: Why not look, for example, at the potential of the Australian "bush potato" (*Ipomoea costata*), well adapted to arid conditions?[63] Like Dahl, Yen is here considering a global use value for indigenous species. But unlike Dahl, and unlike those who sought in earlier times the commodity value of indigenous Pacific plants, he considers the intellectual property rights of the people who have the original cultural connections with the plants in question – reminding us that "benefits from Saccharum canes, rubber from the *Hevea* or *Macadamia* nuts seem not to have reached New Guinea, Amazonia or the Aborigine of Queensland yet."[64]

Studies of areas such as Enga in Papua New Guinea, where traditions of subsistence agriculture remain, and the continuing investigations into the agricultural potential of indigenous plants, can ultimately benefit both individual communities and the wider global community. Attention to past practices can thus be seen to be vitally important to the development of agricultural practices that respect ecosystems and the complex relationship of human beings with the plant world.

Richard H. Grove's *Green Imperialism* looks to the past – to environmental practices of colonial governments – for the seeds of contemporary environmentalism. "Colonial states," he says, "increasingly found conser-

---

to safeguard his social standing, his social claims, his social assets"; Karl Polanyi, *The Great Transformation*, 46.

[63] Yen, "The Origins of Subsistence Agriculture in Oceania," 10.

[64] "The Origins of Subsistence Agriculture in Oceania," 12.

vationism to their taste and economic advantage."[65] The North was altered by its colonization of the South economically, culturally (as Bernard Smith's *European Vision and the South Pacific* shows) and, Grove argues, environmentally. This early environmentalism is a consequence of "the destructive social and ecological conditions of colonial rule" but also of the "growing European consciousness of natural processes in the tropics and [...] a distinctive awareness of non-European epistemologies of nature."[66] The danger of Grove's deeply researched and interesting book is that the way in which he presents his thesis (of a much "greener" colonialism than we might have imagined) paves the road to complacency about colonization and the environment, particularly where my example of the Pacific islands is concerned. It also prompts complacency regarding where we stand now, looking at the problems and challenges we face – daily new ones, such as biotechnology and its consequences for Third-World agriculture. I greatly admire this book for the way it asks us to seek out the ambiguous areas in colonial practice in relation to the land. But I question the conclusion that "rather than being exclusively a product of European or North American predicaments and philosophies, [modern environmentalism] emerged as a direct response to the destructive social and ecological conditions of colonial rule."[67] This is still an utterly eurocentric view: either now or then, the West produced it and there is still no gesture towards what the *other* cultures – the colonized – may have to teach us. It is "our" problem, not "theirs".

Geertz saw anthropological geography stuck in a groove over two questions: environmental impact on culture and the reverse. I see Grove's work as evincing an inability to escape the 'either/or' of condemnation or recuperation of colonialism. He falls back on what he calls irony when really he is stumped by paradoxes. In an article on "The Origins of Environmentalism" he had told us: "Ironically, this new [environmental] sensitivity developed as a product of the specific, and ecologically destructive, conditions of the commercial expansion of the Dutch and English East India Companies."[68] In *Green Imperialism*, too, he has recourse to "irony" to

---

[65] Richard H. Grove, *Green Imperialism: Colonial Expansion, Tropical Island Edens and the Origins of Environmentalism, 1600–1860* (Cambridge: Cambridge UP, 1995): 15.

[66] Grove, *Green Imperialism*, 486.

[67] *Green Imperialism*, 486.

[68] Grove, "The Origins of Environmentalism," *Nature* (3 May 1990): 12.

explain the "conservationist role" of the colonial state.[69] We are not far, here, from Melville's shrug about the guava bush. Grove's shrug – his failure to probe what he terms irony – leaves his book, with all its rich detail, wide open to the accusation of recuperating the colonialist enterprise. The failure to interrogate the yoking of signs of environmental sensitivity with environmental destruction makes this a book addressed solely to the West and in no way to the previously colonized peoples.

At the heart of Grove's study is the "conservation" of Mauritius in the eighteenth century when, he says, the island "became the location for the flowering of a complex and unprecedented environmental policy."[70] Pierre Poivre, "commissaire-intendant" from 1767–72, was responsible for introducing "forest and climate protection measures", establishing a "cordon sanitaire" of timber around the island.[71] Grove links such measures and the development of climate theory to the beginnings of a scientific environmentalism, reflected in the work of people like the naturalist J.R. Forster and the explorer Alexander von Humboldt. We have seen examples of Forster's interest in local environments. Grove, however, cites as an example of early environmental thinking not Forster's observation of indigenous practice, but his comment that seeds of yams, coconut and breadfruit might help to restore the landscape of Easter Island. He relegates to a footnote another of his "ironies" – the fact that "in the same diary entry Forster recommends the introduction of goats and sheep to the Marquesas and New Caledonia as a way of increasing their prosperity – precisely the thinking which had ruined St Helena and played a part in the deforestation of Easter Island."[72] We should be careful about making Forster into a proto-environmentalist. He, too, would "transport" landscapes at will, the imposition of the "useful" being one of the forces driving his imagination about these new worlds, as his list of useful transplantations and introductions shows.[73] But we can value Forster for his ability to listen to and to perceive other ways.

---

[69] Grove, *Green Imperialism*, 7.

[70] *Green Imperialism*, 168.

[71] *Green Imperialism*, 170.

[72] *Green Imperialism*, 327–28; Forster, *The Resolution Journal*, vol. 4: 652.

[73] Forster, *The Resolution Journal*, vol. 4: 652–53. Forster's *Resolution* journal pre-dates Malthus's *Essay on the Principle of Population* (1798) by twenty years or so. But human science, as well as natural science, lurks in his account.

Pierre Poivre was operating in Mauritius in the eighteenth century. And that, too, was when Forster and other naturalists made their astounding discoveries of a new world of plants in the South Seas. The colonization of the Pacific islands was a nineteenth-century phenomenon. As Ellis's programme for the people of Huahine attests, industrial capitalism was running on an extraordinary confidence, abetted by religious certitude. Ellis's drive to make a fallen people industrious mixes very nicely with the demands of a capitalist Empire. What Grove does not show us in his "green imperialism" is the swiftness and carelessness with which the invading Europeans transformed societies and landscapes in the Pacific. The role of natural scientists on exploration expeditions is an interesting one and needs to be more fully examined for its glimpses of Pacific agriculture. But for all their influence in drawing to European attention non-European "epistemologies of nature," their science, too, underwrote irrevocable transformations, as Forster's remarks on Easter Island show.[74] Mauritius had experienced more than a hundred years of colonization when Poivre took his measures. Had those involved in the sandalwood trade learned nothing from the experiences of longer colonized places, such as the islands of the Caribbean? It seems not. Just as it is today, greed was a powerful motivator; and just as it does today, the plunder of nature required a lesser investment. It is still cheaper and easier to cut down old growth forest than to create timber plantations.[75]

Grove generalizes about "colonial governments", but there seems to be little in common between a British government in India having to face up to the consequences of its long rule in a large country with a great many

---

[74] Easter Island was, of course, a "barren" landscape when Forster saw it because of activities of the indigenous population. My point is the too easy assumption that introducing species will automatically solve problems. This thought process is still with us. For every concern about an introduced species gone "wild" – plant or animal – there is talk of introducing another one: witness the recent plan to "unleash" Kenyan wasps in north Queensland to destroy insects damaging citrus crops; *The Australian* (26 September 1999): 22.

[75] Cf Karl Marx: "Natural elements entering as agents into production, and which cost nothing, no matter what role they play in production, do not enter as components of capital, but as a free gift of Nature to capital"; *Capital*, 3: 745, cited in Howard L. Parsons, ed. and comp. *Marx and Engels on Ecology* (Westport CT: Greenwood, 1977): 171.

resources, and the colonial governments – French, British and American[76] – of the islands of the Pacific, whose indigenous populations, compared with those of the older and more slowly colonized islands and continents, hardly knew what hit them.

Grove's study is, I would argue, deeply un-ecological in its failure to examine environments as sets of relationships. It looks at how various colonial governments were manoeuvred into conservation practices, often out of what we could describe, from our twenty-first-century perspective, as genuine concern on the part of some of the colonizers. But Grove does not consider what any of this meant for the indigenous populations. He does not see, any more than Ellis, the intertwined physical and social picture.

When Geertz, in 1963, examined traditional agricultural practices in Indonesia and traced what became of them under Dutch colonial rule, his concern was better to understand Indonesia today. Politics, he taught us, is embedded in farming. For the sake of the Pacific islands and islanders and for the sake of the earth that we now conceive of in its entirety we need to look to the past, and to examine past practices as a way of ensuring that there is debate about current ones. Rather than lament loss, which really digs us deeper and deeper into the groove of our own Western practice, we need to understand what the past, in places like the Pacific islands, can teach us about other ways of living with the land.

WORKS CITED

Aalsberg, W.G.L., & Susan Parkinson. "Traditional Pacific Food Technology," in *Fauna, Flora, Food and Medicine*, ed. Morrison et al., 1–6.

Anderson, Charles R. *Melville in the South Seas* (New York: Columbia UP, 1939).

Brand, Donald D. "The Sweet Potato: an Exercise in Methodology," in *Man Across the Sea*, ed. Riley et al., 343–65.

Brookfield, H.C., & Paula Brown. *Struggle for Land: Agriculture and Group Territories Among the Chimbu of the New Guinea Highlands* (Melbourne: Oxford UP, 1963).

Buck, Peter H. (Te Rangi Hiroa). *Vikings of the Sunrise* (New York: Frederick A. Stokes, 1938).

[76] Hawai'i was titularly an independent kingdom, but was in fact run by the Americans.

Burkill, L.H. *A Dictionary of Economic Products of the Malay Peninsula* (London: Publication on behalf of the Governments of the Straits Settlements and Federated Malay States by the Crown Agents for the Colonies, 1935).

Carlquist, Sherwin. *Hawaii: A Natural History* (New York: American Museum of Natural History, 1970).

Certeau, Michel de. *Heterologies: Discourse on the Other*, tr. Brian Massumi (Minneapolis: U of Minnesota P, 1986).

Crosby, Alfred W. *Ecological Imperialism* (Cambridge: Cambridge UP, 1986).

——. *Germs, Seeds and Animals: Studies in Ecological History* (Armonk NY: M.E. Sharpe, 1994).

Dahl, Arthur L. *Regional Ecosystems Survey of the South Pacific Area* (Noumea, New Caledonia: South Pacific Commission, 1980).

Degner, Otto. *Naturalist's South Pacific Expedition: Fiji* (Honolulu: Paradise of the Pacific, 1949).

Dening, Greg. *Islands and Beaches: Discourse on a Silent Land; Marquesas 1774–1880* (Chicago: Dorsey Press, 1980).

——. *Mr Bligh's Bad Language: Passion, Power and Theatre on the Bounty* (Cambridge: Cambridge UP, 1992).

Denoon, Donald, & Catherine Snowden, ed. *A Time to Plant and a Time to Uproot: A History of Agriculture in Papua New Guinea* ([NP]: Institute of Papua New Guinea Studies, 1981).

Ellis, William. *Polynesian Researches*, 3 vols. (New York: J.J. Harper, 1833).

Forster, J.R. *The Resolution Journal of Johann Reinhold Forster 1772–1775*, ed. Michael E. Hoare, 4 vols. (London: Hakluyt Society, 1982).

Fosberg, F.R., ed. *Man's Place in the Island Ecosystem* (Honolulu: Bishop Museum Press, 1965).

Geertz, Clifford. *Agricultural Involution: the Processes of Ecological Change in Indonesia* (Berkeley: U of California P, 1963).

Gill, W. Wyatt. *Jottings From the Pacific* (London: Religious Tract Society, 1885).

Glacken, Clarence. *Traces on the Rhodian Shore: Nature and Culture in Western Thought From Ancient Times to the End of the 18th Century* (Berkeley: U of California P, 1973).

Golson, Jack. "From Horticulture to Agriculture in New Guinea," in *Historical Ecology in the Pacific Islands*, ed. Kirch & Hunt, 39–50.

Griffiths, Tom, & Libby Robin, ed. *Ecology and Empire: Environmental History of Settler Societies* (Melbourne: Melbourne UP, 1997).

Grove, Richard H. *Green Imperialism: Colonial Expansion, Tropical Island Edens and the Origins of Environmentalism, 1600–1860* (Cambridge: Cambridge UP, 1995).

——. "The Origins of Environmentalism," *Nature* (3 May 1990): 11–14.

Harris, David R. "Agricultural Systems, Ecosystems and the Origins of Agriculture," in *The Domestication and Exploitation of Plants and Animals*, ed. Peter J. Ucko & G.W. Dimbleby (London: Gerald Duckworth, 1969): 3–15.

Hill, Johnson D., & Walter E. Stuermann. *Roots in the Soil: An Introduction to Philosophy of Agriculture* (New York: Philosophical Library, 1964).

Hyams, Edward. *Soil and Civilization* (London: Thames & Hudson, 1952).

Jackson, John R. *Commercial Botany of the Nineteenth Century* (London: Cassell, 1890).

Kirch, Patrick V., & Terry L. Hunt. *Historical Ecology in the Pacific Islands: Prehistoric Environmental and Landscape Change* (New Haven CT: Yale UP, 1997).

Lakau, Andrew A. "Customary Land Tenure and Alienation of Customary Land Rights Among the Kaina, Enga Province, Papua New Guinea" (unpublished doctoral dissertation, University of Queensland, 1994).

Malinowski, Bronislaw. *Coral Gardens and their Magic*, 3 vols. (1935; London: Allen & Unwin, 1966).

Marsh, G.P. *Man and Nature, or Physical Geography as Modified by Human Action* (1864; Cambridge MA: Belknap/Harvard UP, 1965).

Melville, Herman. *Omoo* (1847; New York: Hendricks House, 1969).

Melville, Herman. *Typee* (1846; Chicago: Northwestern UP and the Newberry Library, 1968).

Merchant, Carolyn. *The Death of Nature: Women, Ecology and the Scientific Revolution* (San Francisco: Harper & Row, 1980).

——. *Ecological Revolutions: Nature, Gender and Science in New England* (Chapel Hill: U of North Carolina P, 1989).

Merrill, Elmer D. *Plant Life of the Pacific World* (New York: Macmillan, 1945).

Mitchell, Andrew. *A Fragile Paradise: Nature and Man in the Pacific* (London: Collins, 1989).

Morrison, John, Paul Geraghty & Linda Crowl. *Fauna, Flora, Food and Medicine*, vol. 3 of *Science of Pacific Island Peoples* (Suva, Fiji: Institute of Pacific Studies, 1994).

Nash, Roderick. *Wilderness and the American Mind* (New Haven CT: Yale UP, 1973).

Parkes, Annette. "Environmental Change and the Impact of Polynesian Colonization: Sedimentary Records from Central Polynesia," in *Historical Ecology in the Pacific Islands*, ed. Kirch & Hunt, 166–99.

Parsons, Howard L., ed. & comp., *Marx and Engels on Ecology* (Westport CT: Greenwood, 1977).

Polanyi, Karl. *The Great Transformation* (1944; Boston MA: Beacon, 1957).

Price, A. Grenfell. *The Western Invasions of the Pacific and its Continents* (Oxford: Clarendon, 1963).

Riley, Carroll L. et al., ed. *Man Across the Sea: Problems in Pre-Columbian Contacts* (Austin: U of Texas P, 1971).

Rose, Frederick G.G. *The Traditional Mode of Production of the Australian Aborigines* (Sydney: Angus & Robertson, 1987).

Sahlins, Marshall. *Stone Age Economics* (Chicago: Aldine Atherton, 1972).

Sauer, Jonathan D. "Reevaluation of the Coconut," in *Man Across the Sea*, ed. Riley et al., 309–27.

Shineberg, Dorothy. *They Came for Sandalwood: A Study of the Sandalwood Trade in the South-West Pacific 1830–1865* (Melbourne: Melbourne UP & London: Cambridge UP, 1967).

Simmons, I.G. *Changing the Face of the Earth: Culture, Environment, History* (Oxford: Blackwell, 1989).

Smith, Bernard. *European Vision and the South Pacific* (Oxford: Clarendon, 1960).

Stewart, Charles S. *A Visit to the South Seas in the United States Ship "Vincennes", During the Years 1829 & 1830* (1831; New York: Frederick J. Praeger, 1970).

Waddell, Eric. *The Mound Builders: Agricultural Practices, Environment and Society in the Central Highlands of New Guinea* (Seattle: U of Washington P, 1972).

Walter, Annie. "Knowledge for Survival: Traditional Tree Farming in Vanuatu," in *Fauna, Flora, Food and Medicine*, ed. Morrison et al., 189–200.

Whitney, Henry M. *The Hawaiian Guide Book for Travellers* (Honolulu: Henry M. Whitney, 1875).

Yen, D.E. "Distribution of the Sweet Potato," in *Man Across the Sea*, ed. Riley et al., 328–42.

——. "The Origins of Subsistence Agriculture in Oceania and the Potentials for Future Tropical Food Crops," *Economic Botany* 47.1 (1993): 3–14.

The "I" in Beaver

Sympathetic Identification and Self-Representation in Grey Owl's *Pilgrims of the Wild*

CARRIE DAWSON

IN HER 1972 BOOK *SURVIVAL*, Margaret Atwood compares American animal stories with Canadian animal stories and argues that the former are about killing animals while the latter are about "animals being killed, as felt emotionally from inside the fur and feathers."[1] James Polk made a similar argument in the same year, suggesting that the difference between the two national traditions is that the former is concerned with "the lives of the hunters" and the latter with "the lives of the hunted." More specifically, Polk cites Ernest Thompson Seton's *The Biography of a Grizzly* as an example of a Canadian animal story and proposes that Seton's "sympathetic identification" with his subject is such that readers "have no choice but to identify with his persecuted bears."[2] In an attempt to rejuvenate a critical discussion about self-representation in semi-autobiographical Canadian animal stories, this analysis extends and interrogates the idea of sympathetic identification. But rather than arguing, like Polk and Atwood, that sympathetic identification with persecuted animals is an expression of the Canadian preoccupation with victimhood, I mean to consider the beha-

[1] Margaret Atwood, *Survival: A Thematic Guide to Canadian Literature* (Toronto: Anansi, 1972): 74.

[2] James Polk, "Lives of the Hunted," *Canadian Literature* 53 (Summer 1972): 54, 53.

viours that sympathetic identification produces and the interests that it serves. The second aim of this analysis is to address a concern which Bill Plumstead voiced in a recent letter to the editors of *Maclean's* magazine. Writing in response to an article concerned with the depiction of Grey Owl in Richard Attenborough's eponymously named movie,[3] Plumstead points out that "the Canadian literary establishment" has appraised the nature writer's identity claims while neglecting his writing because a "fraud writing about beavers is apparently an embarrassment to the custodians of CanLit."[4] Because I am sympathetic to Plumstead's argument that literary criticism "is surely an appraisal of texts, not personalities,"[5] I accept his letter as a challenge and, through an analysis of the relationship between sympathetic identification and cross-cultural identification in Grey Owl's *Pilgrims of the Wild* (1935), I will explore one way in which those of us inclined to identify as "custodians of CanLit" might usefully approach the question of how Canadian literary criticism has negotiated – and failed to negotiate – the relationship between and among human and nonhuman cultures.

The "I" in Sympathy

In *The Theory of Moral Sentiments* (1759), Adam Smith defines sympathy as an imaginative act in which we become "in some measure the same person" as the object of our sympathy.[6] Like Smith, the psychologist Robert Katz foregrounds ideas of transformation and coalescence in his more recent definition of sympathy as an "as-if behaviour."[7] But unlike Smith, Katz emphasizes that sympathy involves a degree of self–other differentiation when he writes: "Abandoning our own self, we *seem to* become fused with and absorbed in the inner experience of the other person."[8] According to Katz, sympathy involves an *illusion* of oneness rather than a fusion of self and

[3] Plumstead was responding to Brian D. Johnson's "Rediscovering Grey Owl," *Maclean's* 112.40 (1999): 52–56.

[4] Bill Plumstead, letter, *Maclean's* 112.42 (1999): 4.

[5] Bill Plumstead, letter, *Maclean's*, 4.

[6] Adam Smith, *The Theory of Moral Sentiments*, ed. D.D. Raphael & A.L. Macfie (1759; Indianapolis IN: Liberty Classics, 1982): 9.

[7] Robert Katz, quoted in Thomas J. McCarthy, *Relations of Sympathy: The Writer and the Reader in British Romanticism* (Aldershot, UK: Scolar Press, 1997): 12.

[8] Katz, quoted in McCarthy, *Relations of Sympathy*, 12 (my emphasis).

Other because the self-abandon experienced by the sympathizing subject stops short of total self-abnegation. That said, sympathy is nevertheless *experienced* as a sort of fusion of self and other because the 'as-if' identification with another being temporarily obscures the self–Other differentiation on which the experience of sympathetic identification is predicated.

Although Katz addresses the sympathy that exists between two people, his argument can usefully be applied to the sympathy expressed by authors who identify with their animal subjects. Specifically, Katz's emphasis on the temporary self-forgetfulness of sympathy has important ramifications for the study of nature writing and the subgenre of animal stories, insofar as those who write about the natural world typically downplay their own self-consciousness in an effort to communicate the sympathy and wonder with which they view their surroundings.[9] Because the frequently "self-forgetful" narrators of animal stories tend to communicate their sympathy for the animals about which they write by emphasizing a similarity or kinship between themselves and their subjects, Katz's argument about the assumption of oneness is also of interest. At the very least, it urges us to examine the ways in which oneness gets assumed in animal stories – to ask, "How is the narrator's self-transformation structured?" and "What are its implications?"

Following Thoreau's description of nature writing as a literature which urges the natural world to "flower in a truth" through "direct intercourse and sympathy,"[10] numerous ecocritics have examined the rhetoric of sympathy in nature writing. Relatively few, however, have considered the relationship between sympathetic identification and cross-cultural identification. In the recent history of nature writing in Canada, the need to do so has been made clear on at least three occasions. First, in 1903 the highly regarded American naturalist John Burroughs published an article, "Real and Sham Natural History," in which he argued that claims made by Ernest Thompson Seton in *Wild Animals I Have Known* were so outlandish that the book should

---

[9] See, for example, Scott Slovic, who, as part of his analysis of Annie Dillard's "effacement of authorial identity," argues that "consciousness of the present self interferes with both mystical vision and artistic creation"; Slovic, *Seeking Awareness in American Nature Writing: Henry Thoreau, Annie Dillard, Edward Abbey, Wendell Berry, Barry Lopez* (Salt Lake City: U of Utah P, 1992): 66, 64.

[10] Henry David Thoreau, "Natural History of Massachusetts," in Thoreau, *The Natural History Essays* (Salt Lake City: Peregrine Smith, 1984): 130, 131.

have been called "*Wild Animals I Alone Have Known*."[11] Thirty-five years later, a small Ontario newspaper published an article alleging that Grey Owl, the best-selling author of stories about the Canadian wilderness, had misrepresented himself as a Native American. Moreover, in 1996 *Saturday Night* ran an article by John Goddard in which Goddard argued that the accounts of Inuit culture in Farley Mowat's *People of the Deer* are based on a series of wholly fictitious visits to Inuit camps.

Although Mowat likely romanticized his relationships with Inuit communities in the Canadian Arctic, and Seton identified with Native American cultures to the extent that he often referred to himself as "Black Wolf," it is Grey Owl's identification *as* a Native American that makes him the obvious candidate for a discussion of the relationship between sympathetic identification and self-representation in Canadian animal stories. Thus, in an attempt to determine how the rhetoric of sympathy functions as a vehicle for self-transformation in Canadian animal stories, I will consider Grey Owl's account of his conversion from trapping to animal conservation in his best-selling book *Pilgrims of the Wild.*

The "I" in Beaver

Grey Owl was born Archibald Stansfeld Belaney in 1888 in Sussex, England. In 1906 he left England and moved to Canada, where he worked as a trapper and a guide. While living in the Temagami district of Ontario and later, while serving with the Canadian Armed Forces in France during World War I, Grey Owl circulated false stories about his upbringing in the American south-west as the child of an Apache woman and a Scotsman employed by the Buffalo Bill Show. After returning to Canada, Grey Owl gave up trapping and, at the suggestion of his third wife, Anahareo, he directed his energies toward protecting the beaver. As Grey Owl, he also embarked on a career as a writer and a public speaker. His success was such that, in a bid to increase the profile of the Canadian National Parks, Parks Canada employed

---

[11] John Burroughs, "Real and Sham Natural History," *Atlantic Monthly* 91 (March 1903): 298. See Betty Keller's account of the "Nature Fakers" debate in Chapter 12 of her book *Black Wolf: The Life of Ernest Thompson Seton* (Vancouver & Toronto: Douglas & McIntyre, 1984). Also see Lawrence Buell's short but engaging discussion of Burrough's article in ch. 6 of *The Environmental Imagination: Thoreau, Nature Writing, and the Formation of American Culture* (Cambridge MA: Harvard UP, 1995).

him as the resident conservationist in Saskatchewan's Prince Albert National Park and installed him in a tailor-made cabin called "Beaver Lodge," where he remained until his death in April 1938. Although numerous people suspected that Grey Owl's claims to Native ancestry were exaggerated if not entirely false, their suspicions were not given a public voice until the day after his death, when the *North Bay Nugget* published an article alleging that his identity claims were false. In the following weeks papers all over Canada and England carried the news that he whom Lloyd Roberts had dubbed "the first Indian that really looked like an Indian" was an Englishman.[12]

In his 1973 biography of Grey Owl, Lovat Dickson – who was Grey Owl's publisher and dedicated promoter – described his subject's effect on the Canadian public this way:

> This voice from the forests momentarily released us from some spell. In contrast with Hitler's screaming, ranting voice, and the remorseless clang of modern technology, Grey Owl's words evoked an unforgettable charm, lighting in our minds the vision of a cool, quiet place, where men and animals lived in love and trust together.[13]

On the one hand, Dickson, who was raised in Canada, suggests that the huge attraction that Grey Owl had for the Canadian public can be explained by the resurgence of primitivism in the years leading up to World War II. While recognizing that Grey Owl was an international rather than an exclusively national phenomenon, Dickson, on the other hand, indicates that Grey Owl appealed to Canadians because he provided them (us) with a means of proving the rule of racial difference while attesting to the specificity of Canadian culture. That is, Grey Owl evoked a capitulation to an idealized image of Native difference. He rewarded an audience desirous of a peaceable kingdom by offering them an image of themselves as the benevolent subjects of a beneficent nationstate wherein the appreciation of indigenous culture was such that all Canadians could live "in love and trust together."

Sixty-five years later, our concerns about the destruction of our natural environment and the extinction of an alarming number of nonhuman species has led an ever-increasing number of people to turn to aboriginal

---

[12] Quoted in Daniel Francis, *The Imaginary Indian: The Image of the Indian in Canadian Culture* (Vancouver: Arsenal Pulp, 1992): 131.

[13] Lovat Dickson, *Wilderness Man: The Strange Story of Grey Owl* (Toronto: Macmillan, 1973): 5.

cultures for alternative ways of imagining the relationship between humans and the natural environment. Given the extent to which our anxieties about the welfare of the natural environment have been translated into or expressed as renewed appetite for the platitudes of romantic primitivism (witness the present popularity of T-shirts, tattoos and tourist brochures that employ indigenous iconography), it is not surprising that Grey Owl has recently been resurrected as something of a popular hero. Since 1990 he has been the subject of two biographies, one long poem, numerous essays, and a major – albeit unsuccessful – Hollywood film.[14] In contrast with Grey Owl's early biographers, recent critics are less inclined to naturalize or otherwise rationalize his claims to Native American ancestry, but they do partake of an established consensus that his achievements as a conservationist are ultimately more important than his identity claims. For example, in her 1991 essay "The Grey Owl Syndrome," Margaret Atwood felt compelled to apologize for "drop[ping] the unfortunate Grey Owl" into contemporary debates about identity, authenticity and aboriginality.[15] While I am of the opinion that Grey Owl's long-standing misrepresentation of himself as a Native American is enough to warrant "drop[ping]" him into debates about identity and aboriginality, I share Atwood's reluctance to emphasize his identity claims at the expense of his very successful attempts to popularize the Canadian wilderness. But because Grey Owl's representation of himself is inextricable from his representation of the wilderness, I think that we need to consider how the two projects are related.

Pilgrims of the Wild is a semi-autobiographical account of Grey Owl's conversion from trapper to conservationist and caretaker of beaver. It is largely concerned with the relationship that he and Anahareo had with the beavers who shared their home. Although Grey Owl condenses some major events in his life and omits others,[16] the text is, for the most part, a faithful record

14 The biographies alluded to are Donald B. Smith's *From the Land of the Shadows: The Making of Grey Owl* (Saskatoon: Western Producer Books, 1990) and Jane Billinghurst's *Grey Owl: The Many Faces of Archie Belaney* (Vancouver: Greystone Books, 1999). The long poem, which is written in the form of a biography, is Armand Garnet Ruffo's *Grey Owl: The Mystery of Archie Belaney.*

15 Atwood, "The Grey Owl Syndrome," in Atwood, *Strange Things: The Malevolent North in Canadian Literature* (Oxford: Clarendon, 1995): 37.

16 For example, Grey Owl condenses the time he spent in Saskatchewan's Prince Albert National Park and Manitoba's Riding Mountain National Park by referring only to one unnamed national park.

of Grey Owl's experiences between 1927 and 1934. For the most part, Grey Owl enlists standard autobiographical techniques and refers to himself using the first person, thus narrowing the distance between author and narrator. However, in the first pages of the Preface, Grey Owl maintains a distance between himself and his narrator by using the third person. The Preface begins:

> This is primarily an animal story; it is also the story of two people, and their struggle to emerge from the chaos into which the failure of the fur trade, and the breaking down of the old proprietary system of hunting grounds plunged the Indian people, and not a few whites, during the last two decades.[17]

A few lines later these two people are further described as "a man and a woman [...] [who] broke loose from their surroundings taking with them all that was left to them of the once vast heritage of their people, – their equipment and two small animals as pets" (xiii). A few pages later Grey Owl switches to first-person narration and identifies the "man" and "woman" as himself and Anahareo, who is, he emphasizes, of Mohawk descent. But before doing so, he intimates that "the vast heritage" of which he speaks is Native ancestry. He does not say so explicitly, but he does caution the reader that "it is necessary to remember that though [this] is not an altogether an Indian story, it has an Indian background" (xiii). Grey Owl subsequently explains that the ambiguous phrase "Indian background" signifies an "attitude towards all nature" (xiii). He defines that "attitude" by quoting a passage from John Gifford's "Story of the Seminole War" in which Gifford represents the indigenous people as "a part of nature" (xiv). After expressing his support for Gifford's argument, Grey Owl extends that argument to include "those of other races who have resided for many years in the wilderness" (xiv). The effect of this very inclusive and experientially oriented definition of indigeneity, one which figures indigenous intimacy with nature as a form of training rather than a culturally specific mode of spirituality, is that it allows Grey Owl to establish that he is indigenous without making any genealogical claims.

After establishing his "Indian background," Grey Owl turns his attention to the animal kingdom and attempts to demonstrate the similarly unique nature of his kinship with his animal subjects. First of all, Grey Owl carefully differentiates himself from authors who "ascribe human attributes to

[17] Grey Owl, *Pilgrims of the Wild* (1935; London: Peter Davies, 1939): xiii. Further page references are in the main text.

animals" (xv–xvi) and he argues that if the "qualities" of the animals in *Pilgrims* "are found to closely approximate some of our own, it is because they have, unknown to us, always possessed them, and the fault lies in our not having discovered sooner that these characteristics were not after all exclusively human" (xvi). While declarations of kinship are, of course, standard fare in nature writing, Grey Owl's declarations of interspecies commonality are nevertheless intriguing in light of his subsequent suggestion that the recognition of kinship is reciprocated by the beaver. Specifically, Grey Owl, who came to be known as the "beaver man,"[18] suggests that the beavers attribute animal characteristics to Anahareo and himself. Referring to Jelly Roll – the beaver who became famous in her own right as the star of five films produced by the Canadian Parks Board – Grey Owl says at various points in the narrative: "she took me as much for granted as if I had also been a beaver" (192); "[she] gave me a response of which I had not thought an animal capable" (197); and "[she] seem[ed] to look on me as a contemporary, accepting me as an equal and no more" (198). By suggesting that it is the beaver and not the writer who is transformed by the feeling of sympathy, Grey Owl is able to detract attention from his efforts to create and maintain an illusion of oneness between himself and his subjects. At the same time, he is able to put the onus for "self-forgetful" behaviour on the animals.

If we recall Atwood's and Polk's arguments that Canadian writers of animal stories express a sympathetic identification with their subjects, and William Katz's definition of sympathy as an "as-if behaviour" which involves an illusion of oneness, it is perhaps not altogether surprising that Grey Owl's attempt to communicate his sympathy for the beaver involves identifying *as* a beaver: having stated his desire to avoid employing "coldly formal references to the writer himself in the third person" (185), Grey Owl declares his determination to include "a few good healthy unequivocating 'I's' standing up honestly on their own hind legs" (185). Given that Grey Owl frequently comments on the beavers' predilection for standing on their hind legs and chattering, it can be inferred that he is identifying his narrator with or as a beaver. This is also true of a subsequent episode in which he accompanies his description of returning to his cabin to find that his "quarters have been invaded" with a drawing of an open-mouthed beaver who stands in the doorway of a cabin above a caption that reads, "My quarters have been invaded" (273, 272). Given the placement of the caption, the

---

[18] Jane Billinghurst, *Grey Owl: The Many Faces of Archie Belaney*, 71.

repetition of its contents, and the preceding representation of the first-person narrator as four-legged, the caption can be simultaneously attributed to the beaver, the narrator, and the narrator-as-beaver.

Though it might be argued that Grey Owl's metaphoric representation of a four-legged narrator is somewhat unusual, there is, of course, nothing unusual about the extensive use of metaphors to figure the natural world. As James Olney points out, there is also nothing uncommon about the use of metaphors to figure the self in autobiography. As Olney elaborates,

> A metaphor, then, [is that] through which we stamp our own image on the face of nature, [it] allows us to connect the known of ourselves to the unknown of the world, and, making available new relational patterns, it simultaneously organizes the self into a new and richer entity; so that the old known self is joined to and transformed into the new, the heretofore unknown, self.[19]

Not only does Grey Owl's metaphor "stamp [his] own image on the face of nature," it also takes the narrator's "sympathetic identification" with his subject to its logical conclusion: with the image of a four-legged narrator, Grey Owl, himself, is "transformed" into "a new and richer entity" who is, as the title suggests, "of the wild."

McGinnis and McGinty were the names of the first pair of beaver kits that Grey Owl and Anahareo raised. A year later the kits went missing and the couple adopted two more beavers who came to be known as Jelly Roll and Rawhide. Despite Grey Owl's promise not to "draw comparisons between man and beast, save in a few instances which are too remarkable to be overlooked" (xv), he persistently refers to the beavers with metaphors that liken them to humans. For example, he calls them "Beaver People" (54), "Little Indians" (42), "Immigrants" (93), and "small ambassadors" (129). Jelly Roll is typically referred to as the "Queen" and, after Rawhide loses a piece of his scalp in an accident, Grey Owl takes to calling him "the Little Iroquois" (77).[20] In an article entitled "The Beaver as Native and as Colonist," Gordon Sayre offers one explanation for the wide-ranging and unlikely combination of epithets employed by Grey Owl. Addressing the representation of beavers in Canada between the seventeenth and nineteenth centuries, Sayre explains that the beaver's "double status – as a natural and

[19] James Olney, *Metaphors of the Self: The Meaning of Autobiography* (Princeton NJ: Princeton UP, 1992): 31–32.

[20] He also calls them "queer diminutive Buddhas" (52) and "little folk from some other planet" (53).

social marvel when alive, and as a valuable commodity when dead – caused representations of the beaver to become overdetermined with strong and contradictory ascriptions."[21] By the latter part of the nineteenth century, one of the most common of these ascriptions involved the application of "vanishing Indian" and related "Noble Savage" tropes to the endangered beaver.[22] Like representations of the Noble Savage, idealized portrayals of the beaver were intended to "compensate for or justify its imminent demise."[23] However, the tropes functioned differently in at least one important way: the rhetoric of Noble Savagery underscored *differences* between indigenous and colonial culture, while the figurative construction of the "Noble Beaver" underscored *likenesses* between the animal's habits and early colonial culture in Canada.[24] Eighteenth-century French travel writers like Pierre Charlevoix and Nicolas Denys used the beaver to "elaborate a vision of colonial society with all the hierarchy and social control of France."[25]

Like Charlevoix and Denys, Grey Owl used the beaver to articulate a vision of an ideal society. However, Grey Owl's "vision" differed from that of Charlevoix and Denys in at least two ways. First, and most importantly, Grey Owl's ideal society was predicated on a respect for animal rights. Secondly, while Charlevoix and Denys depended on a dichotomized representation of beavers and Natives to articulate their vision, Grey Owl's idealized vision of the "wilderness" assumed an essential continuity between beavers and Native peoples. By sympathetically identifying with both, Grey Owl was able to naturalize his production of himself as Native: beavers, wrote Grey Owl, "were of the Wild as we were" and "we," he added, "are Indian" (163).

Although Grey Owl certainly valued indigenous cultures as idealized images of life close to nature, he also valued them for their ability to inter-

---

[21] Gordon Sayre, "The Beaver as Native and as Colonist," *Canadian Review of Comparative Literature/Revue Canadienne de Littérature Comparée* 22 (septembre/décembre 1995): 659.

[22] Sayre, "The Beaver as Native and as Colonist," 660.

[23] "The Beaver as Native and as Colonist," 665.

[24] Sayre demonstrates that numerous eighteenth- and early-nineteenth-century French travel writers and naturalists who visited Canada responded to French concerns that early settlers would mimic the lifestyle of the indigenous people by representing the "cooperative, industrious, and non-nomadic beaver [...] as a model for colonists to imitate"; "The Beaver as Native and as Colonist," 660.

[25] Sayre, "The Beaver as Native and as Colonist," 671.

pret, use and express the natural environment. This is evident in *Men of the Last Frontier*, where he argues that "[Indians] have catalogued and docketed every possible combination of shape, sound, and colour possible in their surroundings," so "a disarrangement in the set of leaves; the frayed edge of a newly broken stick, speak loud to the Indian's eye."[26] To the extent that this excerpt demonstrates Grey Owl's respect for the "the Indian" as animist and archivist, it also suggests that his use of a metamorphosing narrator in *Pilgrims* constitutes an allusion to an aboriginal tradition of storytelling in which changes between animals and humans occur frequently and are integral to the story. The problem is, however, that even as Grey Owl underscores the virtues of an indigenous narrative tradition given to interpreting and expressing an environmental consciousness, he unwittingly demonstrates his distance from such a tradition by representing his relation with the nonhuman realm as largely a solitary affair: despite the narrating "I" who "stand[s] up honestly on [his] own hind legs" in order to produce a narrator that is very much at home in the wilderness, images of isolation and alienation persist throughout *Pilgrims*. In the last pages of the book Grey Owl underscores the absence of a community that shares his ecological consciousness. He declares, for example, that "the pilgrimage is over,"[27] but it is evident that his "pilgrimage," unlike most, has not ended with a homecoming. The beavers are hibernating and, Anahareo, unable to put up with Grey Owl's solitary life-style, has left. Thus the writer pictures himself alone in the cabin which Parks Canada constructed as an exact replica of one that he had shared with Anahareo and the beavers. "Atavistic?" he asks. "Perhaps it is, but good has come of it" (281). Though good certainly did come of Grey Owl's very successful attempt to found a beaver colony, his effort to represent that process exhibits a weakness common to the genre that Atwood calls "realistic animal stories."[28] In the Introduction to *Family of Earth and Sky*, John Elder and Hertha Wong identify that weakness very elegantly by explaining that these narratives have helped to heal "the rift between the 'two cultures' of science and literature, [...] [but that] they have also projected the voices of solitary – and sometimes alienated individuals rather more often than they have em-

[26] Grey Owl, *Men of the Last Frontier* (1931; Toronto: Macmillan, 1989): 219.

[27] Grey Owl, *Pilgrims of the Wild*, 282.

[28] Atwood,, "The Grey Owl Syndrome," 73.

phasized how the human community might be seen as part of nature."[29] Following Wong, Elder, and Robert Harrison – who points out that human beings, unlike other species, "live not in nature but in relation to nature"[30] – we might conclude that the problem with some Canadian animal stories is that the sympathetic identification which produces an "I" who lives in nature has failed to produce a "we" who live "in relation" to nature.

The "I" in Nature

Because Grey Owl "live[d] in relation to nature," he faced the problem of how to represent himself "in nature" in writing. And because the self-conscious act of writing about nature is at odds with the self-forgetful wonder with which the nature writer approaches his subject, he had to enlist a number of strategies to get the writing out of nature writing. Thus Grey Owl represents himself as a very unlikely author whose fingers are "stiffened a little by the paddle and the pull of a loaded toboggan" and are consequently "ill suited" to what he calls the "writing game" (xvi). He then elaborates on his supposed lack of experience with the English language by casting aspersions on the "factory-made English" that he claims to have very recently retrieved from the "cold storage where it had languished for the better part of three decades" (200). As Donald Smith notes, Grey Owl's characterization of himself as someone who has only recently acquired literacy belies his position as the top student in English at Hastings Grammar School in Sussex, England.[31] More importantly, it also belies the self-consciousness with which he crafted his sentences so as to appear like "factory-made English."[32]

In the latter part of *Pilgrims of the Wild*, Grey Owl reminisces about the process of writing his earlier book, *The Men of the Last Frontier*. He says that

---

[29] *Family of Earth and Sky: Indigenous Tales of Nature from Around the World*, ed. John Elder & Hertha Wong (Boston MA: Beacon, 1994): 3.

[30] Robert P. Harrison, "Towards a Philosophy of Nature," in *Uncommon Ground: Toward Reinventing Nature*, ed. William Cronon (New York: W.W. Norton, 1995): 426.

[31] Donald B. Smith, *From the Land of the Shadows: The Making of Grey Owl* (Saskatoon, Saskatchewan: Western Producer, 1990): 19.

[32] Donald Smith points out that one way Belaney made his English appear "factory-made" was to stipulate that his publishers could not correct any of the numerous grammatical errors or unconventional sentence structures in his work; Smith, *From the Land of the Shadows*, 118.

he had "no idea how to put a book together" when he embarked on that project. He writes:

> I bethought me of the 'Writing System' that Anahareo had brought away from her home thinking they were cookery books. I dug them out of the oblivion in which they had so long lain forgotten and soon became deeply absorbed in matters of "Setting," "Dialogue," "Point of View," "Unity of Impression," and "Style." (199)

The result, says Grey Owl, is that his "stories seemed to have the peculiar faculty of writing themselves, quite against any previous plans [he] made for them" (200). With this characterization of his stories as self-writing, Grey Owl effectively disavows self-consciousness, strategy and agency. By disavowing self-consciousness, he also renounces the tension that is inherent in the genre of nature writing – namely, the tension between the transcendental wonder which nature evokes and the self-consciousness involved in representing that wonder. Having done so, he is able to claim that he "never felt so close to [Nature]" as he did when writing about it (203).

Like Grey Owl, the American nature writer Annie Dillard proposes that the process of writing can be performed in obliviousness to the writing subject. In *The Writing Life*, a collection of reflections on writing, Dillard instructs her readers that "process is nothing; erase your tracks."[33] Because I have devoted a considerable part of this essay to exploring the various ways in which Grey Owl erased his "tracks," I should add that there are relatively few points in the text where he does so by making direct claims to indigenous ancestry. Declarations like the aforementioned "for we are Indian" are rare (163). More frequent are slightly ambiguous references to the author's tendency temporarily to "rever[t] to the savagery of forgotten ancestors" (246).[34] Likewise, Grey Owl writes that "people were kind to a buckskin clad sauvage, and [...] a woman of a conquered race" (171). As Judith Dudar has noted, the difference between Grey Owl's characterization of himself and his characterization of Anahareo in this passage is instructive because, rather than including himself among the "conquered race," Grey Owl employs the

---

33 Annie Dillard, *The Writing Life* (New York: Harper & Row, 1989): 4.

34 When, for example, Grey Owl writes "And ever in my heart there was an aching loneliness for the simple kindly people, companions and mentors of my younger days, whose ways had become my ways, and their gods, my gods," it is unclear whether he refers to the Apache with whom he claimed to have grown up, the Ojibway with whom he later affiliated himself, or woodsmen and rural people of mixed race.

more ambiguous "sauvage," a French word that sounds like "savage" but translates as "wild."[35] While Dudar is right to point out that passages of this kind "allow misinterpretation,"[36] it is also worth noting that the passages which allude to or affirm Grey Owl's indigeneity are often attributed to others. For example, he writes that "there were not a few who looked askance at [us] passing Indians" (82), and he recalls meeting a French-Canadian couple who appear friendly but say little to him because they assume – or so Grey Owl thinks – that he spoke "only Indian" (74). By representing himself as he imagines he is seen by others, Grey Owl is able to construct his self-representation as consensual while also putting the onus for identificatory claims on others.

In addition to the intradiegetic audience who corroborate Grey Owl's identity claims there was, of course, a sizeable extradiegetic audience who did the same thing. In part, they did so because Grey Owl successfully exploited a widespread appetite for nostalgic images of – to borrow a phrase – "imaginary Indians."[37] He also exploited an equally widespread willingess to subsume the "Native" under the category of the natural, to assume that indigenous people are "of the wild," that they are authorities on all aspects of "the wild," and that they, like "the wild," are quickly disappearing. In numerous books, magazine articles, films and lectures, Grey Owl exploited these primitivist fantasies in order to sell his own fantasy of a place where "the scars of ancient firs are slowly healing over [...] and the beaver towns are filling up again" (282). If we consider that Grey Owl attended over 200 meetings and addressed over 500,000 people in the four months between November 1935 and February 1936,[38] we can, no doubt, agree that his success was as spectacular as his beaded buckskin get-up. Given this, it strikes me that the best strategy is to exploit his present popularity. Perhaps the thing to do is to use Hollywood's depiction of Grey Owl as an opportunity to consider how aboriginal culture is spectacularized in Canada today. Equally, at a time when his books are being republished, we might consider

---

[35] Judith Dudar, "Feint of Heart and Sleight of Hand: Autobiographical Art and Artifice in the Life-Stories of Will James, Grey Owl, and Frederick Phillip Grove" (unpublished doctoral dissertation, Dalhousie University, Canada, 1994): 176.

[36] Dudar, "Feint of Heart and Sleight of Hand," 173.

[37] In *The Imaginary Indian*, Daniel Francis argues that "the Indian is the invention of the European" and is thus "anything non-Natives wanted [him] to be" (4–5).

[38] Jane Billinghurst, *Grey Owl: The Many Faces of Archie Belaney*, 101.

how his representation of himself as a pilgrim "of the wild" perpetuates an idea of the natural world as an isolated "pristine landscape that we ourselves do not inhabit."[39] And finally, perhaps we might heed the advice of another author of semi-autobiographical animal stories whose own identity claims have recently been subject to scrutiny – namely, Farley Mowat. When the journalist John Goddard questioned Mowat about the veracity of his published accounts of the time he spent in the Canadian Arctic, he is said to have exclaimed, "I never let the facts get in the way of the truth!" and, somewhat less poetically, "Fuck the facts!"[40] Elsewhere, Mowat has offered a somewhat more subtle explanation for the sorts of inconsistencies identified by Goddard. In the Preface to a catalogue of his papers at McMaster University, Mowat characterizes himself as "a teller of tales."[41] By way of conclusion, I want to suggest that we follow Farley Mowat and Bill Plumstead by paying close attention to the imaginative, rhetorical and otherwise *literary* dimensions of animal stories. In respect to works by Grey Owl, this involves foregoing sensational accounts of assumed identity, and replacing them with more provocative questions about the relationship between sympathetic identification and strategic self-representation in a genre of writing where the former has been consistently over-determined and the latter has been typically under-emphasized. It seems to me that if we endeavour to think systematically about what Canadian animal stories and Canadian literary criticism have to say to one another, we may well be able to use romantic tales about nature writers living "in nature" to teach us what it is to live ethically and responsibly in "relation to nature."[42]

---

[39] William Cronon, "The Trouble with Wilderness; or, Getting Back to the Wrong Nature," in *Uncommon Ground: Toward Reinventing Nature*, ed. William Cronon (New York: W.W. Norton, 1995): 88.

[40] Quoted in John Goddard, "A Real Whopper," *Saturday Night* 111.4 (1996): 49.

[41] Quoted in Goddard, "A Real Whopper," 49.

[42] I am grateful to the Social Sciences and Humanities Research Council of Canada for supporting my research.

WORKS CITED

Atwood, Margaret. "The Grey Owl Syndrome," in Atwood, *Strange Things: The Malevolent North in Canadian Literature* (Oxford: Clarendon, 1995): 35–61.

——. *Survival: A Thematic Guide to Canadian Literature* (Toronto: Anansi, 1972).

Billinghurst, Jane. *Grey Owl: The Many Faces of Archie Belaney* (Vancouver: Greystone, 1999).

Buell, Lawrence. *The Environmental Imagination: Thoreau, Nature Writing, and the Formation of American Culture* (Cambridge MA: Harvard UP, 1995).

Burroughs, John. "Real and Sham Natural History," *Atlantic Monthly* 91 (March 1903): 298–309.

Cronon, William. "The Trouble with Wilderness; or, Getting Back to the Wrong Nature," in *Uncommon Ground: Toward Reinventing Nature*, ed. William Cronon (New York: W.W. Norton, 1995): 69–90.

Dickson, Lovat. *Wilderness Man: The Strange Story of Grey Owl* (Toronto: Macmillan, 1973).

Dillard, Annie. *The Writing Life* (New York: Harper & Row, 1989).

Dudar, Judith. "Feint of Heart and Sleight of Hand: Autobiographical Art and Artifice in the Life-Stories of Will James, Grey Owl, and Frederick Phillip Grove" (unpublished doctoral dissertation, Dalhousie University, Canada, 1994).

Elder, John, & Hertha D. Wong, ed. *Family of Earth and Sky: Indigenous Tales of Nature from Around the World* (Boston MA: Beacon, 1994).

Francis, Daniel. *The Imaginary Indian: The Image of the Indian in Canadian Culture* (Vancouver: Arsenal Pulp, 1992).

Goddard, John. "A Real Whopper," *Saturday Night* 111.4 (1996): 46–50, 52, 54, 64.

Grey Owl [aka WA-SHA-QUON-ASIN]. *Men of the Last Frontier* (1931; Toronto: Macmillan, 1989).

——. *Pilgrims of the Wild* (1935; London: Peter Davies, 1939).

Harrison, Robert P. "Towards a Philosophy of Nature," in *Uncommon Ground: Toward Reinventing Nature*, ed. William Cronon (New York: W.W. Norton, 1995): 426–37.

Johnson, Brian D. "Rediscovering Grey Owl," *Maclean's* 112.40 (1999): 52–56.

Keller, Betty. *Black Wolf: The Life of Ernest Thompson Seton* (Vancouver & Toronto: Douglas & McIntyre, 1984).

McCarthy, Thomas J. *Relations of Sympathy: The Writer and the Reader in British Romanticism* (Aldershot, UK: Scolar Press, 1997).

Olney, James. *Metaphors of the Self: The Meaning of Autobiography* (Princeton NJ: Princeton UP, 1992).

Plumstead, Bill. Letter, *Maclean's* 112.42 (1999): 4.

Polk, James. "Lives of the Hunted," *Canadian Literature* 53 (Summer 1972): 51–59.

Sayre, Gordon. "The Beaver as Native and as Colonist," *Canadian Review of Comparative Literature/Revue Canadienne de Littérature Comparée* 22 (septembre/décembre 1995): 659–82.

Slovic, Scott. *Seeking Awareness in American Nature Writing: Henry Thoreau, Annie Dillard, Edward Abbey, Wendell Berry, Barry Lopez* (Salt Lake City: U of Utah P, 1992).

Smith, Adam. *The Theory of Moral Sentiments*, ed. D.D. Raphael & A.L. Macfie (1759; Indianapolis IN: Liberty Classics, 1982).

Smith, Donald B. *From the Land of the Shadows: The Making of Grey Owl* (Saskatoon: Western Producer, 1990).

Thoreau, Henry David. "Natural History of Massachusetts," in Thoreau, *The Natural History Essays* (Salt Lake City: Peregrine Smith, 1984): 1–29.

The Sandline Mercenaries Affair

Postcoloniality, Globalization and the Nation-State*

ROBERT DIXON

IN A RECENT ARTICLE on the global trade in body parts, John Frow takes "postcolonial theory" to task for its apparent inability to explain the forces of globalization. "The focus of postcolonial theory," Frow argues, "[. . .] is on the aftermath of empire"; reasoning from evidence about the trade in body parts, he suggests that "ties of dependence in the contemporary world are not primarily ties between nations." The focus of postcolonial theory on "relations between an emergent nation-state and the former imperial nation-state mean[s] that [it] has largely failed to come to terms with the major explanatory accounts of these multi-layered ties of dependence. [...] Increasingly, cultural dependencies follow the routes of capital flows through global capital markets."[1]

In the present discussion I want to test Frow's argument through an analysis of the Sandline mercenaries affair in Papua New Guinea (PNG) in 1996–97. In examining the affair, I follow Frow's injunction to draw on theoretical perspectives derived from world-systems theory, anthropology and international studies. Such models also have the advantage of supplementing literary postcolonialism's strong disciplinary emphasis on texts and

* For information and advice during the preparation of this essay, I am grateful to Richard Gehrmann, Philip Kitley, Christopher Lee, Meaghan Morris, Brian Musgrove, Glen Ross and Helen Tiffin.

[1] John Frow, "Private Parts: Body Organs in Global Trade," *UTS Review* 1.2 (1995): 99.

discourses with an analysis of political and economic factors. My main concern, however, is to read these two bodies of theory dialectically, rather than as competing or mutually exclusive paradigms. Frow's sense of the word "aftermath," as I understand it here, is like Lyotard's "post": it implies continuity and renewal as much as, or even more than, the sense of an ending. My point is not the negative one, that postcolonial theory cannot explain globalization, but the more positive one, that globalization is still profoundly affected by what Frow calls "the aftermath of empire."

Postcoloniality and Globalization

John Frow is not, of course, the first critic to point out the limitations arising from postcolonial theory's retaining the nation-state as a key actor in its explanatory models. In an article significantly titled "Postcolonialism and Globalisation," Simon During also warns that "economic structures and flows are tied less and less tightly to national economies."[2] Literary postcolonialism, in particular, in both its theorization and its curriculum, has tended to be organized around a narrative of colonial expansion by European nation-states followed by nationalist resistance and liberation. The empire, conceived here as a group of postcolonial nation-states, "writes back" to the centre.[3] Paul Sharrad has argued that this "national focus [...] has excluded consideration of wider or more subtle links, including the increasingly diffused economic and communications system controls of publishing and cultural dissemination." Sharrad also remarks that its growing interest in diasporas has meant that "Postcolonial studies as a set of inter-national and regional practices is already straining at its definitional boundaries." To cope with this, the discipline "requires a more complex [...] theory."[4]

In a survey of futures for postcolonial studies, Patrick Williams cites a range of theorists whose work might usefully contribute to such a project, particularly sociologists associated with theories of world systems and

---

[2] Simon During, "Postcolonialism and Globalisation," *Meanjin* 2 (1992): 342.

[3] Cf. Bill Ashcroft, Gareth Griffiths & Helen Tiffin, *The Empire Writes Back* (London: Methuen, 1989): 2.

[4] Paul Sharrad, "'One Size Fits All?': Internationalisation and Postcolonial Studies," in *Globalisation and Regional Communities: Geoeconomic, Sociocultural and Security Implications for Australia*, ed. Donald McMillen (Darling Heights: U of Southern Queensland P, 1997): 245–46.

globalization: they include Immanuel Wallerstein, Roland Robertson, Bryan Turner, Anthony Giddens, Mike Featherstone and Arjun Appadurai.[5] The point of such arguments for postcolonial theory is that postcolonialism must move beyond explanatory models based solely on the nation-state, and on the interactions between nation-states grounded in such binaries as centre–periphery and colonizer–colonized. In an influential formulation of this argument, Arjun Appadurai argues:

> The new global cultural economy has to be seen as a complex, overlapping, disjunctive order, which cannot any longer be understood in terms of existing centre–periphery models (even those which might account for multiple centres and peripheries). The complexity of the current global economy has to do with certain fundamental disjunctures between economy, culture and politics which we have barely begun to theorise.[6]

Appadurai's model of this economy is based on five global flows or vectors. First, *ethnoscapes* produced by flows of people: tourists, immigrants, refugees, exiles, guestworkers. Second, there are *technoscapes*, the machinery and plant flows produced by multinational and national corporations and government agencies. Third, there are *finanscapes*, produced by the rapid flows of money in the currency markets and stock exchanges. Fourth, there are *mediascapes*, the repertoires of images and information, the flows which are produced and distributed by newspapers, magazines, television, film and the Internet. Fifth, there are *ideoscapes*, linked to flows of images which are associated with state or counter-state movement ideologies, comprised of elements of the Western Enlightenment world-view: images of democracy, freedom, welfare and rights. Such a model of disjunctive global flows might be used to analyse the Sandline mercenaries affair in a more nuanced way than earlier post-colonial theories have allowed. It also has the advantage of paralleling recent developments in the writing of Pacific history, which show an increasing awareness of cultural hybridization and exchange, and the radical ambiguity of all claims to ethical agency.[7]

---

[5] Patrick Williams, "No Direction Home? Futures for Postcolonial Studies," *Wasafiri* 23 (1996): 3–6.

[6] Appadurai, "Disjuncture and Difference," 296.

[7] See Nicholas Thomas, "Fear and Loathing in the Postcolonial Pacific," *Meanjin* 2 (1992): 271–72. These issues are discussed further in his article "Partial Texts: Representation, Colonialism and Agency in Pacific History," *Journal of Pacific History* 25 (1990): 139–58.

The "black hole at the heart of Bougainville"[8]

The origins of the Sandline mercenaries affair lie in the dispute over the operation of Conzinc Riotinto of Australia's (CRA's) copper mine at Panguna, and the unequal agency of the participants in this venture – the local landowners, the regional government, the state of Papua New Guinea and a transnational mining corporation. The history and operation of the mine illustrate the complex relations between postcoloniality and globalization. As John Connell and Richard Howitt note in their study of mining and indigenous peoples in the Pacific, "the intersecting interests of mining companies and indigenous peoples are woven into complex webs of social, political, and economic relations, which shape development opportunities and constraints for each group."[9] Far from representing a break with the past, however, this postcolonial present is constituted by the legacy of colonialism, whose problems it often compounds. Despite the technical dispossession of native peoples by colonial laws, for example, many remote peoples were never effectively dispossessed of their lands in the colonial era. Capital, after all, allows for and encourages "private" ownership; it fetishizes "real estate." It does not necessitate the expulsion of native peoples, but their possessions, their land, their life-ways, are only meaningful when cashed in, in the terms of the capitalist system of exchange and value. Globalization, therefore, might have its paradigm in imperial capitalism, and may not be as "untheorizable" as Appadurai assumes – there may already be a model for it in the theory of Capital. But the former remoteness of these people has left some areas as "storehouses of mineral wealth accessible to a new era of exploration and extraction technologies, and new forms of economic and juridical organization in the form of transnational resource companies and [newly] independent [national] governments."[10] Many of these new states, formed as a consequence of decolonization, have gone out of their way to attract transnational corporations, often waiving environmental protection laws and overriding the interests of their own regional and minority groups.

---

[8] Cf. Andrew Marshall, "Behind the Lines," *Weekend Australian Magazine* (5–6 April 1997): 21–26.

[9] John Connell & Richard Howitt, "Mining, Dispossession, and Development," in *Mining and Indigenous Peoples in Australasia*, ed. Connell & Howitt (Melbourne: Oxford UP/ Sydney UP, 1991): 1.

[10] Connell & Howitt, "Mining, Dispossession, and Development," 5.

Panguna is typical of this pattern in the South West Pacific. The mining company CRA began exploration on Bougainville in 1964. When copper was discovered in 1967, Bougainville Copper Limited (BCL) was established, two-thirds owned by CRA and one-third by Broken Hill Consolidated. BCL was granted leases for mining, tailings disposal, road access and town and port facilities. These concessions were granted under the Mining Ordinance of 1928, a legal instrument of the colonial era that did not require an Environmental Impact Survey. The mine began production in 1972 and quickly became one of the largest in the world. Since 1972, BCL has contributed about 16 percent of PNG's internally generated income and 44 percent of its exports. But the distribution of profits from the mine over the ten year period to 1987 gives a dramatic indication of the disjunction between the interested parties: 60 percent went to the national government of PNG as taxes, fees and dividends; 35 percent to foreign shareholders; 5 percent to the North Solomons Provincial Government; and 0.2 percent went to local landowners in royalty payments.[11]

Concern over the distribution of profits from the mine fuelled secessionist sentiments on Bougainville, culminating on 1 September 1975 in a widely supported declaration of independence as the Republic of the North Solomons. The secessionist movement stressed the distinctive cultural identity of the Bougainville people and the necessity to achieve a greater share of the mine profits. In 1988, after a long period of dissatisfaction, the Panguna Landowners' Association finally called on BCL to pay it 50 percent of all profits and K10 billion compensation for long-term environmental damage. Dissatisfied with BCL's response, villagers blocked access roads, and in November 1988 militant landowners bombed mine installations, a situation that led to the closure of the mine in May 1989. Francis Ona, the Commander of the BRA, declared an independent Republic of Bougainville in May 1990. The PNG government responded by cutting all shipping and telecommunications links, and imposed a blockade to stop supplies reaching the island.[12]

CRA's relationship with Bougainville certainly exemplifies the artificiality of the postcolonial nation-state and the conduct of transnational companies,

---

[11] John Connell, "Compensation and Conflict: The Bougainville Copper Mine, Papua New Guinea," in *Mining and Indigenous Peoples in Australasia*, ed. John Connell & Richard Howitt (Melbourne: Oxford UP / Sydney UP, 1991): 55.

[12] Connell, "Compensation and Conflict," 55–56.

but the Nasioi people, on whose land the mine was built, also view the environmental destruction and diversion of wealth away from their local region as a continuation of colonialism. The Bougainville Freedom Movement's Internet site, managed from Erskineville in Sydney, uses the virtual space of the world wide web to publicize the secessionist cause, but its homepage harks back to the cartographic practices of the colonial era: "At the turn of the century, power-broking European colonial powers made Bougainville part of Papua New Guinea (PNG) despite the fact that Bougainville is 500 kilometres away from the mainland."[13] One Nasioi man explains his understanding of colonial history in this way:

> When my grandfather was alive and my father was a little boy, the Germans came. They gave us steel axes and cloth to cover our bodies. Then the Australians chased away the Germans and the Japanese chased away the Australians. The Americans chased away the Japanese so the Australians could come back. Now my grandfather is dead, my father is old, and I am a man. And what do we have? Still nothing more than steel axes and sarongs.[14]

These examples are more than a simple misrecognition of the capital flows of postmodernity. They also indicate an accurate appreciation of how those flows have been facilitated by the structural relations of imperial and postcolonial nation-states.

Dogs of War

When Sir Julius Chan regained Prime Ministership of PNG in August 1994, he promised to end the secessionist war on Bougainville. He signed a ceasefire with the BRA and organized a peace conference which produced the Bougainville Transitional Government in 1995. But a peace settlement with the BRA in January 1996 collapsed almost as soon as it was reached, and rebel attacks on PNG soldiers intensified. As a consequence, the government of Sir Julius Chan abandoned peace talks, and in March 1996 renewed the military push into Central Bougainville in Operation High Speed. Far from finishing off the BRA, however, Operation High Speed was a victory

---

[13] Bougainville Freedom Movement Homepage: http://www.magna.com.au/~sashab/ BFM/ background.html

[14] Quoted in Eugene Ogan, "Living Among the Nasioi of New Guinea," in *Contemporary Cultural Anthropology*, ed. Michael Howard (Glenview IL: Scott, Foresman, 3rd ed. 1988): 63.

for the rebels, who followed up their success by killing twelve soldiers and capturing another five.[15]

On 22 February 1997, the *Australian*'s South-Pacific correspondent, Mary–Louise O'Callaghan, broke the news that PNG had secretly engaged a group of foreign mercenaries to assassinate the rebel leadership on Bougainville and reopen the Panguna mine. PNG dealt directly with London-based Sandline International, who subcontracted the Pretoria-based Executive Outcomes to provide personnel. PNG's Finance Minister, Chris Haiveta, directed the Bank of PNG to transfer K33.6 million to a company called Roadco, which was the payment vehicle for the mercenaries. The money was not part of the legitimate 1997 budget appropriations, and the indirect payment was an attempt to conceal the contract from the International Monetary Fund and the World Bank, as it constituted a violation of PNG's current loan agreements with those bodies. By mid-February, forty mercenaries were established in a training camp at Urimo, 35 kilometres from Wewak.

In Australia, the Howard government vigorously condemned the arrangement and, frustrated by its stand-off with PNG, put a promised $4 million relief package on hold. In this instance, Howard was clearly playing the colonial relationship to suit his own purposes, for in pronouncing Sandline's involvement as "totally unacceptable," he neglected to mention that the Australian government had itself for many years fulfilled a similar role through its military assistance to PNG, the most notable recent consequence of which was the controversial use of Australian-donated helicopter gunships in Bougainville. In March, "secret" talks were held between Howard and Chan at Kirribilli House in Sydney. Faced with the threat that Australia might withdraw its $320 million annual aid program to PNG, Chan announced that the Sandline staff would not now be allowed on Bougainville, and that their role would be solely advisory. But within days Chan rejected Howard's pressures and defiantly asserted that the mercenaries were indeed to be used on Bougainville. Then on 17 March, in what has come to be called the St Patrick's Day mutiny, the Commander of the PNGDF, Brigadier-General Jerry Singirok, called for the resignation of the Prime Minister, Sir Julius Chan, together with his Deputy and Defence Minister, over his government's "devious actions" in the Sandline affair. In Operation

---

[15] Stewart Firth, "Leader Rode Back to Power on Promise to Settle Dispute," *Australian* (19 March 1997): 6.

Rausim Kwik, Singirok ordered PNGDF personnel not to serve with Sandline staff and sealed all military installations, including those on Wewak, effectively placing the Sandline personnel under arrest in Moem Barracks. Sandline CEO Tim Spicer was captured in a covert operation in Port Moresby and placed under military arrest. In response, Chan called an emergency cabinet meeting, removed Singirok from office, and appointed Colonel Alfred Aikung as Acting Commander. But on 19 March, as Howard sent Australia's most senior diplomat, Philip Flood, to Port Moresby as his "personal emissary," soldiers and civilians protested in the streets, demanding that Chan resign and that Singirok be reinstated as Commander of the PNGDF. On 20 March, in response to the Australian delegation's threat to withdraw Australian aid and the pressure of the demonstrations, Chan called an extraordinary meeting of cabinet at which the contract with Sandline was suspended. Justice Warwick Andrew was appointed to lead a commission of inquiry into the contract, whose findings were handed down on 29 May, just days before the elections in which Chan lost government.[16]

The Sandline deal was a complex series of negotiations between Spicer, Chan, Singirok, Haiveta and others, involving a number of public and commercial institutions, and taking place over many months, and in various locations, including Port Moresby, Singapore, Cairns, London, Pretoria and Hong Kong. Evidence given by Tim Spicer to the Andrew Inquiry in April 1997 dramatically confirmed the underlying commercial nature of the contract. When it became apparent during negotiations that PNG's funds might be inadequate, Spicer wrote formally proposing a joint venture in which Sandline and Executive Outcomes were to be paid by both the PNG government and CRA to re-secure the mine and operate it to their mutual benefit. When this letter was tabled at the Andrew Inquiry by Singirok's counsel, Spicer is reported to have "visibly blanched."[17]

Although the "joint venture" appears to confirm arguments that international flows of capital are unconstrained by nation-states, it needs to be remembered that in the Sandline affair as a whole, it is precisely the idea of the nation-state, and its associated legal instruments and ethical authority, that allowed Singirok to mobilize widespread support, and to bring about the events that led to the mercenaries' departure and Chan's standing down.

---

[16] This account is based on the files of the *Australian* newspaper.

[17] Mary–Louise O'Callaghan, "Spicer Blanches under a Grilling," *Australian* (5–6 April 1997): 14.

Singirok and others were able to exploit the ethical crisis arising from Chan's engagement of the mercenaries because of a conflict between the rhetoric of economic rationalism and more traditional ideas about nation-states and their civic responsibilities. The Sandline deal was seen to be unethical because it violated the idea that nation-states are vested with the exclusive right to deploy military force. Through February and March of 1997, numerous articles in the *Australian* tried to make sense of the troubling relationship between soldiering and commerce. In "Men who Put the Sold into Soldier," for example, Joan Beaumont rightly pointed out that the association between soldiers and the nation-state is relatively recent – a feature of modern history since the nineteenth century – as exemplified by the famous inscription in the chapel of the British officer training college at Sandhurst, where Spicer was a cadet: "*Dulce et decorum est pro patria mori.* Sweet and proper it is to die for one's country." Beaumont concludes that "This identification of military service with nationalism meant that those who were prepared to kill in the service of another country – one for which they had no emotion and to which they owed no loyalty – became pariahs."[18]

The strategic ethical value of nationalism can also be measured in the contradictory statements made by Sandline staff during the affair. An article in the *Sydney Morning Herald* described Sandline executives as "speaking the language of the free market with accomplished ease." Tim Spicer is quoted as saying, "I suspect a lot of this hoo-ha would not have arisen if the people who are running the training were still in the uniform of a national army." "All Sandline is doing," he asserted, "is filling a void in the market to meet the needs of a client." Yet, when pressed, even Spicer sought refuge in the ethics of the nation-state, pointing out: "We are a contractor to a Government that is the internationally recognized and democratically elected Government of PNG and has the right to determine what is necessary for its internal needs."[19]

The continuing authority of the nation-state as an ethical basis for violence was reinforced in a Report on the question of the use of mercenaries compiled at this time by the United Nations Commission on Human Rights,

---

[18] Joan Beaumont, "Men who Put the Sold into Soldier," *Australian* (26 February 1997): 13.

[19] Christopher Henning, "PNG's British Forces," *Sydney Morning Herald* (27 February 1997): 2.

which includes an account of the situation in PNG.[20] The report places a high value on national sovereignty as the ethical basis of military power, and disputes the justification offered by Spicer both to the press and to the Andrew Inquiry – that is, that this company was employed by and therefore answerable to a sovereign state:

> … this is one of the most controversial aspects of the issue, particularly because no Government is authorised to exercise the attributes of authority against the sovereignty of the State itself, but also because responsibility for internal order and security in a sovereign country is an obligation which may not be renounced or transferred and which the State discharges through its police and armed forces. The State may privatise many things and many services that lie within its competence, but not that which constitutes its very *raison d'être*. If it hands over such authority to a private company, and a foreign private company at that, it is agreeing to a limitation of State sovereignty.[21]

The Colonial Relationship

In one sense then, the Sandline affair reveals that the state of PNG was an artefact of colonialism, imperfectly imagined, its sovereignty vulnerable to internal secession and to the flows of international capital. Yet those forces were constrained by arguments and legislation arising from PNG's status as a nation-state and a former colonial territory. I want to look now at two ways in which the aftermath of empire continues to shape discourse: first in the complex game of brinkmanship between Chan and Howard; and second, in the representation of the affair in key elements of the Australian media, especially in News Limited's the *Australian.*

In the days immediately following public exposure of the contract, the Chan government was able to gain support within PNG by mobilizing the anti-Australian sentiment that was a legacy of colonialism. Chan's complex game with the Australian government involved appealing to anticolonial sentiments in PNG just long enough to give him some purchase against

---

20 *Report on the question of the use of mercenaries as a means of violating human rights and impeding the exercise of the right of peoples to self-determination, submitted by Mr. Enrique Bernales Ballestros, Special Rapporteur, pursuant to Commission resolution 1995/5 and Commission decision 1997/120.* Published by the United Nations Commission on Human Rights, 27 January 1998. http://www.unhchr.ch/ html/menu4/chrrep/ 98chr3. html

21 *Report on the question of the use of mercenaries*, para 74.

Howard's interventions without losing the benefits of Australia's continuing colonial obligations in the form of aid. This might be compared with Malaysian Prime Minister Dr Mahathir Mohamad's playing of the "culture card" by invoking the idea of "Asian values" against Australia during the *Embassy* and "Recalcitrant" Affairs of 1991 and 1993. The effect, as Peter Searle argues, was to mobilize the idea of Malaysian nationalism and "to perpetuate a false dichotomy between Australia and the region that permitted only conditional [Australian] membership of it."[22] In its editorial of 25 February, the PNG *National* suggested that Australia's "excited" reaction and its "badmouthing of PNG" were "because of its fear of losing its influence here."[23] After the revelations, the PNG Opposition closed ranks with Chan against Australia's interventions. The PNG Opposition leader, Roy Yaki, said, "The Australian government has gone too far in its criticisms of a decision made by a government of an independent state." The Opposition frontbencher and Member for Wewak, Bernard Narokobi, claimed that Australia continued to treat PNG "like its colony," which was "totally unacceptable." Narokobi was engaging in some colonial mimicry: the phrase "totally unacceptable" was an insolent echo of Howard's own early response to the use of mercenaries in the South Pacific.[24]

It is also clear that, from Howard's point of view, the discourse of colonialism placed constraints on what he could say in public. When Howard appeared on Channel 10's *Meet the Press*, he acknowledged the limits of what Australia could do:

> It's not the sort of issue where one should make over-the-shoulder comments on the run. [...] You are dealing with a very sensitive area of international relations and something that is right on our doorstep and something that the rest of the world sees Australia as having a particular responsibility for. You're also dealing with an independent country and you have the sensitivity of a former colonial power telling a former colony what it thinks it ought to be.[25]

---

[22] Peter Searle, "Recalcitrant or *Realpolitik*? The Politics of Culture in Australia's Relations with Malaysia," in *Pathways to Asia: The Politics of Engagement*, ed. Richard Robison (Sydney: Allen & Unwin, 1996): 72.

[23] "Nothing mercenary about a better military," *National* (25 February 1997), cited on BFM homepage.

[24] Peter Niesi, "Opposition chiefs leap to government aid," *Post Courier* (27 Saturday 1997), cited on BFM homepage.

[25] Quoted in Sean Dorney, *The Sandline Affair: Politics and Mercenaries and the Bougainville Crisis* (Sydney: ABC, 1998): 252.

Like Howard, Chan was therefore playing the colonial relationship both ways to suit himself. When Howard sought to intervene, Chan invoked PNG's sovereignty against the paternalism of its former colonial master. But when Australia accused him of unethical conduct, he accused Australia of neglecting its fiduciary obligations to a former colony: "I think the Australians have let us down. I just can't imagine really why the Australians are not coming spontaneously to help us bring all this to a conclusion."[26] Invoking the colonial past for one last turn of the screw, the Australian Opposition leader, Kim Beazley, remarked that "The Australian Government has every reason to believe that Sir Julius's nose has been thumbed at it from the outset."[27]

Jerry Singirok and the Uncivil Native Body

In its reporting of the St Patrick's Day mutiny and the ensuing civil unrest, the mainstream Australian media drew heavily on the repertoire of colonial discourse, suggesting that the former colony still needs Australian guidance, and that beneath its veneer of civility PNG is in danger of reverting to a primitive, precolonial past. From the moment he took action against the mercenaries, Singirok was characterized as a modern, postcolonial figure, yet his virtues were always linked to his training in Australia. The *Australian*'s Brian Woodley provides the history of his training and service in Australian institutions, including the Australian Army School of Infantry at Singleton and the Australian Defence Force's Land Warfare Centre at Canungra in Queensland.[28] Mary–Louise O'Callaghan notes that he is "a highly disciplined Australian-trained officer,[29] describing him as "the Army's moral crusader."[30] Immediately after Singirok's sacking, however, speculation began that in his current role, as a private citizen with inside knowledge of

---

[26] "Attempts to play down role of mercenaries in PNG," *Australian Associated Press* (26 February 1997), cited on BFM homepage.

[27] Georgina Windsor, "Howard Condemns Military Rebels," *Australian* (19 March 1997): 6.

[28] Brian Woodley, "Protégé Turns on Sir Julius," *Australian* (18 March 1997): 6.

[29] Mary–Louise O'Callaghan, "General Conducts Mutiny via Radio," *Australian* (18 March 1997): 1.

[30] Mary–Louise O'Callaghan, "Army's Moral Crusader goes Fishing," *Australian* (27 March 1997): 6.

government and the military, he might become "a loose cannon." This suggests something of the ambivalence surrounding Singirok's public persona: his military "discipline" which derives from his association with Australia may be a veneer covering a Melanesian propensity to incivility. Editorials began to hint at this incivility, implicitly comparing the siege of the PNG parliament to a Fijian-style military coup:

> Despite the rightfully sacked Brigadier Singirok's continued assertions that the military were not interested in a coup, the pictures of agitated troops and their *ad hoc* involvement in the siege of Parliament during the past two days have sent shudders of apprehension around the Pacific. It is important that the PNG Defence Force is brought under control by strong leadership that is committed to support the country's democratic processes and acts on the orders of the civil power.[31]

Howard's warnings prior to the popular demonstrations in Port Moresby on 19 March 1997 had also contained veiled implications about the potential breakdown of civil society. His problem was to find the words to imply such a threat without resorting to the patronizing, or worse, racist language of a former colonial master. On the day of Singirok's mutiny Howard announced, "It is this sort of destabilization that we feared the introduction of such mercenaries into PNG and the region might cause."[32] Despite his resistance to Howard's paternalism, Chan too played on this fear of the uncivil native when he criticized Singirok, implying, contrary to reports of his "highly disciplined" character, that the Brigadier-General "went out of his mind a little bit."[33] Insofar as Singirok represents the colonial body, he therefore personifies a conflict between nature and nurture, between civility and incivility, the "highly disciplined" persona he has acquired by being Australian-trained, and his implicitly Melanesian tendency to irrational violence.

This incivility was now to be played out on the streets of Port Moresby and sharply invoked by Australian media reports of the "riots." A photograph in the *Australian* shows a tableau dramatically divided into two parts, with a PNG policeman in immaculate Western-style uniform on the left, disciplining looters on the right. The caption reads, "Long arm of the law [...] police arrive as looters flee Port Moresby shops with their booty yesterday."

---

31 Editorial, *Australian* (27 March 1997): 12.

32 John Short, David Nason & Georgina Windsor, "Concerned Howard Claims Vindication," *Australian* (18 March 1997): 6.

33 Mark Stewart, "General 'out of his mind'," *Australian* (19 March 1997): 1.

Leisa Scott's report from Port Moresby invokes the idea of a collective native body, an image of riot and carnival disrupting the veneer of western civility:

> "Do you know what is happening?" the young man yelled, almost drowned out by the thunder of running feet. "We are going to loot the Chinese stores."
>
> And with that, the wave of civil unrest that has been building in Papua New Guinea since mercenaries arrived earlier this year burst its banks, unleashing a torrent of anger and resentment....
>
> They came from everywhere, hundreds of people – boys, old men – arming themselves with whatever they could pick up and smash against the heavily fortified walls of the Street Scene store....
>
> Police stood over the doors of the store, by now jam-packed with looters, the teargas billowing. When it seemed time they let the doors go, opening the betel-juice stained pavement to the looters who streamed out, still clinging to whatever they had laid their hands on.[34]

In "PNG's Enemies Within," Mary–Louise O'Callaghan revealed that the "riots" had sparked deep-seated insecurities among Port Moresby's middle-class residents. The "enemies within" are poverty and corruption. By implication, they are "native" rather than Western in origin, and there is evidence of regression since independence. As one informant put it, "In the first five years after independence we had [...] mostly Australians who were running departments." Since independence there has been an "erosion of government efficiency." The fear of native incivility is captured in a striking image from the "riots": "a private security guard lopped off a would-be looter's hand, leaving it to fall into a mound of spilt rice."[35]

Underlying all these analyses of the uncivil native body is an assumption that the incivility has its origins in Melanesian rather than in Western culture, which is the source of institutions that can control the threat of regression. In the streets of Port Moresby and on Bougainville, where Western institutions are visibly in ruins, civil society – *westernized* society – is reverting to the primitive. Only Singirok's induction into the military culture of the West allows him to control this. As reporters constantly remark, he is a "disciplined" person.[36]

---

[34] Leisa Scott, "Mutiny and the Bounty," *Australian* (20 March 1997): 4.

[35] Mary–Louise O'Callaghan, "PNG's Enemies Within," *Weekend Australian Review* (19–20 April 1997): 1–2.

[36] Trevor Watson, "On a Knife Edge," *Bulletin* (1 April 1997): 19.

End of Empire?

I began this essay with John Frow's suggestion that "the aftermath of empire" might be read dialectically in relation to processes of globalization. If the complexities of the Sandline mercenaries affair are any indication, the relation between colonialism and postcolonialism is not one of simple succession, for, as Malcolm Crick argues, "cultural processes created under colonial circumstances continue beyond the end of empire." Crick is one of an increasing number of anthropologists critical of the term postcolonialism:

> The attendant risk is that such labels profoundly disguise the many varied ways in which there are deep continuities between "now" and "then". [...] The actors may have changed somewhat [...] but the underlying processes are very similar and the "colonial" dimension is therefore perpetuated. [...] Nation-states in the Third World are therefore frequently just as entangled with metropolitan centres as they were before "independence".[37]

It is a shortcoming of the kind of theoretical models Frow and others advocate for postcolonial studies that they tend to be historically depthless. Far from being displaced by globalization, ideas of the nation-state, of relations between nations that were formerly defined by empire, and even residual fiduciary obligations arising from colonialism, play a prominent and strategic role in the languages of law, international diplomacy and media representation which surround the Sandline affair. Appadurai proclaims that "the new global cultural economy [...] cannot any longer be understood in terms of existing center–periphery models," apparently rejecting the continuity of imperialist dependencies. Yet this is contradicted by his own observation that the relational "flows" of globalization are still "inflected very much by the historical, linguistic and political situatedness of different sorts of actors," including nation-states, multinationals and sub-national groupings.[38] The appropriate model is perhaps not a paradigm shift but a palimpsest. As Masao Miyoshi argues, the ghosts of the colonial past, including the "illusion of the national community," stubbornly persist into the

[37] Malcolm Crick, "Entangled Lives and Meanings: Colonialism and its Cultural Legacy," in *Global Forces, Local Realities: Anthropological Perspectives on Change in the Third World*, ed. Bill Geddes & Malcolm Crick (Geelong, Victoria: Deakin UP, 1997): 66–67.

[38] Appadurai, "Disjuncture and Difference in the Global Cultural Economy," 296.

age of transnational corporations.[39] I would argue that it is to the complex relation between nation-states, still inscribed in legislation and public memory, and the global network of transnational corporations, that postcolonial theory might look to define what Miyoshi characterizes a global "reconfiguration" of colonialism. As Paul Sharrad puts it, we need to develop a body of theory "that still recognises the structuring force of the national while engaging with international movements of labour and post-national constructions of identity and power."[40]

Analysis of the Sandline affair also takes us beyond the debate over whether relations of domination in global politics are primarily transnational or primarily interstate/international. In many ways the case study illustrates how transnational dependency relations interact with and even express themselves through national dependency relations. In particular, the case study seems to contradict any argument that the inter-national and colonial dimensions of global events are merely "residual." To the contrary, they are very much to the fore, interacting with and harnessing transnational flows. The Sandline affair therefore breaks down the false dichotomy between inter-national colonial relations and transnational relations. It could also help in separating colonial / non-colonial distinctions from inter-state / transnational distinctions: the relationship between Sandline and the PNG government is a colonial one, and in many ways the relationships between Australia and the PNG government (mediated through CRA, News Limited and the Internet, for example) are transnational. After all, it was not simply that Sandline was a privatization of PNG's military function, but that here was a realization and exposure of the extent to which the state itself acted as the client of a large corporation – CRA. This is a dimension of the affair which the Howard government, which does the same, would surely not want seen. It adds a layer of complexity to the affair as a chain of client-dependencies, where the state appears not as independent, but as the "middleman" of a complex series of relations that are at once postcolonial and transnational.

But perhaps the last word on the continuing force of colonial relations in a global context should go to Tim Spicer. Reporting on the Sandline affair, Sean Dorney recounts Spicer's conversation with Brian Lowe, the British High

---

[39] Masao Miyoshi, "A Borderless World? From Colonialism to Transnationalism and the Decline of the Nation-State," *Critical Inquiry* 19 Summer (1993): 744.

[40] Paul Sharrad, "'One Size Fits All?': Internationalisation and Postcolonial Studies," 246.

Commissioner in Port Moresby, into whose charge Spicer was released from police custody:

> The police placed six armed policemen around the High Commissioner's residence. Lowe was amused by this because while the military saw it as keeping Spicer in, Spicer was thankful it kept the soldiers out. But the High Commissioner, a Scotsman, was not so amused at Spicer's wit. "The night we got him out, after he'd cleaned up and showered, we were having a gin and tonic on the verandah and I asked, "How do you get into the Scots Guards, Tim? There's nothing Scottish about you." And the cheeky bugger replied, "Well, you know, High Commissioner, all our colonial troops need white officers!"[41]

WORKS CITED

Appadurai, Arjun. "Disjuncture and Difference in the Global Cultural Economy," *Theory, Culture and Society* 7.2–3 (1990): 295–310.

Ashcroft, Bill, Gareth Griffiths & Helen Tiffin. *The Empire Writes Back: Theory and Practice in Post-Colonial Literatures* (London: Methuen, 1989).

Beaumont, Joan. "Men who Put the Sold into Soldier," *Australian* (26 February 1997): 13.

Connell, John. "Compensation and Conflict: The Bougainville Copper Mine, Papua New Guinea," in *Mining and Indigenous Peoples in Australasia*, ed. John Connell & Richard Howitt (Melbourne: Oxford UP/Sydney UP, 1991): 55–75.

——, & Richard Howitt. "Mining, Dispossession, and Development," in *Mining and Indigenous Peoples in Australasia*, ed. Connell & Howitt, 1–17.

Crick, Malcolm. "Entangled Lives and Meanings: Colonialism and its Cultural Legacy," in *Global Forces, Local Realities: Anthropological Perspectives on Change in the Third World*, ed. Bill Geddes & Malcolm Crick (Geelong, Victoria: Deakin UP, 1997): 63–113.

Dorney, Sean. *The Sandline Affair: Politics and Mercenaries and the Bougainville Crisis* (Sydney: ABC, 1998).

During, Simon. "Postcolonialism and Globalisation," *Meanjin* 2 (1992): 339–53.

Firth, Stewart. "Leader Rode Back to Power on Promise to Settle Dispute," *Australian* (19 March 1997): 6.

Frow, John. "Private Parts: Body Organs in Global Trade," *UTS Review* 1.2 (1995): 84–100.

Henning, Christopher. "PNG's British Forces," *Sydney Morning Herald* (27 February 1997): 2.

Marshall, Andrew. "Behind the Lines," *Weekend Australian Magazine* (5–6 April 1997:) 21–26.

---

41 Sean Dorney, *The Sandline Affair: Politics and Mercenaries and the Bougainville Crisis* (Sydney: ABC, 1998): 108.

Miyoshi, Masao. "A Borderless World? From Colonialism to Transnationalism and the Decline of the Nation-State," *Critical Inquiry* 19 Summer (1993): 726–51.

O'Callaghan, Mary–Louise. "Army's Moral Crusader goes Fishing," *Australian* (27 March 1997): 6.

——. "General Conducts Mutiny via Radio," *Australian* (18 March 1997): 1–6.

——. "PNG's Enemies Within," *Weekend Australian Review* (19–20 April 1997): 1–2.

——. "Spicer Blanches under a Grilling," *Australian* (5–6 April 1997): 14.

Ogan, Eugene. "Living Among the Nasioi of New Guinea," in *Contemporary Cultural Anthropology*, ed. Michael Howard (Glenview IL: Scott, Foresman, 3rd ed. 1988): 63–65.

Report on the question of the use of mercenaries as a means of violating human rights and impeding the exercise of the right of peoples to self-determination, submitted by Mr. Enrique Bernales Ballestros, Special Rapporteur, pursuant to Commission resolution 1995/5 and Commission decision 1997/120. Published by the United Nations Commission on Human Rights, 27 January 1998. http://www. unhchr.ch/ htm1/menu4/chrrep/98chr3.html

Scott, Leisa. "Mutiny and the Bounty," *Australian* (20 March 1997): 4.

Searle, Peter. "Recalcitrant or *Realpolitik*? The Politics of Culture in Australia's Relations with Malaysia," in *Pathways to Asia: The Politics of Engagement*, ed. Richard Robison (Sydney: Allen & Unwin, 1996): 56–84.

Sharrad, Paul. "'One Size Fits All?': Internationalisation and Postcolonial Studies," in *Globalisation and Regional Communities: Geoeconomic, Sociocultural and Security Implications for Australia*, ed. Donald McMillen (Darling Heights: U of Southern Queensland P, 1997): 245–54.

Short, John, David Nason & Georgina Windsor. "Concerned Howard Claims Vindication," *Australian* (18 March 1997): 6.

Stewart, Mark. "General 'out of his mind'," *Australian* (19 March 1997): 1.

Thomas, Nicholas. "Fear and Loathing in the Postcolonial Pacific," *Meanjin* 2 (1992): 265–76.

——. "Partial Texts: Representation, Colonialism and Agency in Pacific History," *Journal of Pacific History* 25.2 (December 1990): 139–58.

Watson, Trevor. "On a Knife Edge," *Bulletin* (1 April 1997): 19–22.

Williams, Patrick. "No Direction Home? Futures for Postcolonial Studies," *Wasafiri* 23 (1996): 3–6.

Windsor, Georgina. "Howard Condemns Military Rebels," *Australian* (19 March 1997): 6.

Woodley, Brian. "Protégé Turns on Sir Julius," *Australian* (18 March 1997): 6.

Planting the Seeds of Christianity

Ecological Reform in Nineteenth-Century Polynesian London Missionary Society Stations

ANNA JOHNSTON

IN 1876, THE REV. WILLIAM WYATT GILL published his memoir of missionary work for the London Missionary Society (LMS), *Life in the Southern Isles.* The illustration which accompanied his text (Figure 1) was a crucial summary of the benefits that Gill believed Christianity had brought to the Polynesian region. It constituted a "before and after" narrative of the effects which British missionary intervention was believed to have brought about: the illustration suggests a profound transformation of the physical, built, and moral environment from that which had existed prior to Christian evangelization. It is a visual text rich in signifiers, the most significant of which are those indicating the change in the use of public space (from a stage for "heathen ritual" to a productive village square), a transformation in the self-presentation of individuals within such a space (from nakedness to clothing, leisure to labour, groups to individuals), and land use and productivity (from religious ritual to agricultural and industrial productivity). This discussion demonstrates that each of these issues was central to missionary intervention in the physical spaces of Polynesia in the early nineteenth century.

This intervention and the concomitant questions of ecological, economic and agricultural reform in Polynesian mission stations are here examined through their representation in LMS missionary narratives from the first half of the nineteenth century. This analysis argues that, in attempting to turn Polynesia into what a contemporary historian has described as "a Pacific version of a Birmingham suburb,"[1] LMS missionaries brought about significant and ongoing environmental and social change in Pacific communities. Missionary ecological imperialism was both a part of, yet clearly differentiated from, broader imperial incursions into "new" colonial environments. Significantly, missionary environmental intervention was distinguished by its reliance upon a pre-existing theological doctrine and discourse which facilitated environmental practices.

From the time of early European voyages to the islands of the South Pacific, European interest in what were regarded as "island paradises" was intense. Romantic ideas about the lost Eden, the noble savage, and utopian island cultures pervaded European visions of the area, resulting in acute public attention and a subsequent demand for narrative accounts of these idealized locales. Missionary societies were also energized by this public interest and soon established missions to the region. The LMS's first overseas mission was to what was then called Otaheite (now Tahiti); considerable personnel, funds and interest were invested in this area throughout the nineteenth century. South Pacific missions always had a special place in the history and memory of the LMS.

This analysis concentrates on evangelical activity in the Polynesian region, commonly assumed to include the islands of the central and eastern Pacific, with Hawai'i, Samoa, Tahiti and Tonga, and excluding Micronesia and Melanesia. As Nicholas Thomas has noted of the geographic entity "Melanesia," "Polynesia" is "an artefact of colonial ethnology,"[2] but the term is employed here to establish a sense of congruence across these islands' experience of LMS missionaries.[3] These mission communities were originally established in the LMS's first concerted move into the South Pacific at the end of the eighteenth century.

---

[1] Graeme Kent, *Company of Heaven: Early Missionaries in the South Seas* (Wellington: Reed, 1972): 48–49.

[2] Nicholas Thomas, *Entangled Objects: Exchange, Material Culture, and Colonialism in the Pacific* (Cambridge MA: Harvard UP, 1991): 53.

[3] While I am concerned that this terminology does not conflate the very different cultures and the often disparate missionary encounters experienced in these various islands, it is demonstrable that they experienced the impact of the LMS, its policies, and representatives at a comparable time and in comparable ways; and that this historical and cultural context is crucial to any analysis of missionary activity.

Following the LMS's formation as a non-denominational, Protestant missionary society in 1795, the Directors needed to establish their first overseas mission. The choice of "the South Seas" as the first LMS mission site was somewhat contentious, with at least two of the dominant members of the Directorial committee (Drs Haweis and Bogue) having strongly divergent opinions on the ideal location. The enthusiasm for the "South Seas" is historically seen to be due to the Directors' reading of the published journals and accounts of Captain James Cook, whose description of a kind of "island paradise" caught the imagination of much of the British public of the day. As a result, the LMS established their inaugural mission in Polynesia in 1796. It is significant that textual representation of this new colonial field was influential, through editions of Cook's journals, from the beginning of the Society's involvement there, as it prefigured the prolific publication market which fed (upon) European visions of the region.

The South Pacific grew to fill a crucial role in the LMS's foreign activities and in the British public's social consciousness. Choosing the South Pacific proved to be a shrewd public relations exercise on the part of the LMS, because accounts of the Society's experiences there built upon the already awakened public interest in the area. The first missionaries to Polynesia were sent off with great fanfare and publicity but they were strictly controlled by the Directors of the Society, who were slightly dismissive of their ill-educated missionary representatives and were eager for them not to be awarded hero status. Later, more successful missionaries to Polynesia took hold of the public imagination and became folk heroes in their own right.

The first group of LMS missionaries to Polynesia was singularly unsuccessful in its evangelizing efforts. In retrospect, the 1796 mission was ill-conceived and, like much missionary work, fuelled more by religious zeal than informed planning. While the Society possessed some information about what their envoys might expect to encounter in the islands,[4] individual missionaries seem to have been utterly confounded by the cultural difference they met with and the extreme isolation they experienced. Like the first missions to Aborigines in Australia, the South Seas missions almost all

[4] The LMS gained detailed information about Polynesian society and language from the returned *Bounty* mutineers; William Ellis, *Polynesian Researches, During a Residence of Nearly Six Years in the South Sea Islands; Including Descriptions of the Natural History and Scenery of the Islands – with Remarks on the History, Mythology, Traditions, Government, Arts, Manners, and Customs of the Inhabitants*, 2 vols. (London: Fisher & Jackson, 1829), vol. 1: 71.

thoroughly failed within the first five years and, by the end of the eighteenth century, only seven missionaries were left in the entire South Pacific field.[5]

The second generation of missionaries to Polynesia had much greater success. They built upon the knowledge of those older missionaries, some of whom returned to the field after their initially disappointing experience. John Williams was sent to Ra'iatea (the Society Isles) in 1817 and became one of the LMS's most prominent missionary figures, a kind of David Livingstone of the Pacific. By the mid-1800s, LMS missions in the Pacific were being held up as exemplars of achievement, and in 1830 the Rev. James Sherman was using the South Pacific as inspiration for further missionary efforts. His sermon exhorted:

> Turn to the Islanders of the Southern Ocean, a people among whom were cherished every unhallowed passion, every debasing vice, and every species of cruelty; but by the preaching of the gospel their kings have become nursing fathers, and their queens nursing mothers, to the church – their judges peace, and their exactors righteousness. Agriculture, architecture, mechanics, and ship-building have progressed with astonishing rapidity, and for the rectitude of their laws, the morality of their conduct, and the vigour of their efforts to spread the gospel around them, are at this moment, perhaps, the most righteous and God-fearing nation on the face of the earth.[6]

Like most eighteenth- and nineteenth-century Europeans, LMS missionaries expected the South Pacific to be a kind of "island paradise," and represented it through what Richard Grove calls an "edenic island discourse."[7] However, the Protestant ideology exemplified by the LMS missionaries required the "wanton" fruitfulness of Polynesian environments to be transformed into a Calvinist agricultural economy characterized by commerce and Christianity.

Aignificant part of the Polynesian mission was the introduction of what the missionaries and their backers considered the arts of civilization. In Polynesia, the philosophy of "christianization through civilization" was central to

---

[5] Stephen Neill, *A History of Christian Missions* (The Penguin History of the Church; Harmondsworth: Penguin, 1964): 252.

[6] Rev. James Sherman, *The Remembrance of Christ's Love a Stimulus to Missionary Exertion: A Sermon, Preached before the London Missionary Society, at Spa Field's Chapel, on Thursday, May 13th, 1830* (London: Fisher & Jackson, and Nisbet, n.d.): 21.

[7] Richard Grove, *Green Imperialism: Colonial Expansion, Tropical Island Edens, and the Origins of Environmentalism, 1600–1860* (Studies in Environment and History; Cambridge: Cambridge UP, 1995): 5.

LMS missions from the very beginning. In some ways, the early missionaries focused on "civilization" at the expense of "christianization." Later evangelists were more successful in the integration of the two. Central to the process of "civilization" were attempts to intervene in cultural issues such as the education of children, Polynesian extended-family housing, gender-segregated cultural practices (other than those sanctioned by Christian ethics), agriculture, industry, and housing. The missionaries continually had to negotiate the often competing demands of civilization and christianization. Because they were frequently in the British public's eye thanks to their own Society's publicity machine, the LMS representatives in Polynesia were often criticized for getting the balance wrong.

For the LMS missionaries in Polynesia, the key to effective Christian conversion was always felt to be "civilization," a process of Christian reform which inevitably involved industry and commerce. Missionaries deliberately restructured Polynesian communities through the transformation of both the natural and the built environment – that is, through both agricultural and housing reform. The introduction of agricultural work and European land-use to indigenous Polynesian communities was integral to missionary social reform. Missionaries attempted to establish industries including sugar mills, cotton mills, and coconut oil mills in order to provide productive, Christian labour for their potential native converts. These industries were largely unsuccessful because Polynesian leaders either withheld their consent, or their peoples' labour, from these doomed enterprises. The introduction of European-style housing also became a measure of the success of the LMS missionaries, particularly given the implications that housing reform had on gender and family relations – another critical part of the missionary project. Implicit in the missionary philosophy was an assumption that such material transformations would effect concomitant moral and religious transformations and would produce the basis for a modern, ordered Christian community. As Greg Dening suggests of missionary attempts at transforming the people of the Marquesan islands, the Enata:

> All the missionaries, Catholic and Protestant, English, French and American [...] sensed a relationship between religious ideas and social realities. Enata could never see the grace of God until a different perception of their Land gave them a different philosophy of cause and effect. Their double world was too particular, their gods too familiar to allow acknowledgement of a Cause. They could not be reasoned with until their sense of time and space was remade, until they played

out the realities of a Christian cosmology in the roles and institutions of Aoe's [the European's] world.[8]

As Paul Sharrad has argued, there was an unmistakable "link between agriculture and the civilizing mission of colonialism."[9] Missionaries in Polynesia (and indeed throughout the colonial mission field) constantly used metaphors of "planting the seeds of Christianity" in colonial climates; in "sowing the seed" of Christian messages; and in "cultivating" the development of native Christians. Missionary writers, never reluctant to extend a metaphor beyond its logical breaking point, consistently used these organic metaphors to describe their evangelical work with colonized people.

In these writings, such images did considerable textual and cultural work. They connected the practice of Christianity with, for example, a natural order of things, thereby overwriting a precolonial "state of nature" with a prescriptive and constructed *Christian* "state of nature." Alternative colonial customs were thus seen as "unnatural" and "artificial," and paradoxically needed to be "returned" to an Edenic state through the imposition of Christian values. This kind of contradictory logic characterizes much missionary discourse, and its ambivalence contributes to the interestingly doubled nature of much missionary textuality.

Metaphors of organic Christianity also inscribed the inevitability of Christian conversion, even in unfamiliar colonial climates. They attempted to tie together the advances of science and religion in the imperial exploration of the new world. Missionary writers, particularly in the South Pacific, were keenly interested in scientific discovery and their texts are frequently filled with scientific observation alongside Christian rhetoric. John Williams wrote his *Narrative of Missionary Enterprises in the South Sea Islands* in 1837 and began it with a long discussion of the natural and scientific importance of the region. He wrote, "almost innumerable beauties were found to exist, bestudding the bosom of the vast Pacific with their beauties."[10] Williams is referring here to the islands themselves, though his highly gendered description is significant. In the generalized desire of many Europeans for the regions,

---

[8] Greg Dening, *Islands and Beaches: Discourse on a Silent Land: Marquesas, 1774–1880* (Chicago: Dorsey, 1980): 178.

[9] Paul Sharrad, "Putting Down Roots: Colony as Plantation," *SPAN* 46 (April 1998): 73.

[10] John Williams, *A Narrative of Missionary Enterprises in the South Sea Islands: With Remarks upon the Natural History of the Islands, Origin, Languages, Traditions, and Usages of the Inhabitants* (1837; London: Snow, 1838): 5.

geographical and sexual characteristics often stood in for each other, and the desired conquest of both land and native women was thus made explicit. Williams went on to classify first-, second-, and third-class islands, by way of their geographical features and vegetation.[11] His classificatory system was heavily influenced by European ideas of the sublime – the "first class" are mountainous and volcanic; the second class hilly, with beautiful vegetation but, he wrote, "less sublime their character";[12] and the third are coralline islands. Williams also devoted ten pages to explaining the ecology of coral and coral islands.[13]

For the new breed of "Renaissance-men" missionaries like Williams, this engagement with the emergent scientific community did not contradict religious beliefs – rather, for them it confirmed that there was indeed a logical "master plan" devised by an omniscient God. Paradoxically, given the contemporary opposition of scientific and religious communities, nineteenth-century missionaries were frequently involved as the kind of amateur scientists Mary Louise Pratt discusses in *Imperial Eyes*.[14] As Niel Gunson suggests,

> the "new breed" of missionaries sent to infuse life into the Pacific mission stations of the London Missionary Society immediately following the declaration of peace in Europe in 1815, were supposedly better trained than their predecessors, more practical and less pietistic, more appreciative of knowledge for its own sake, more imbued with a sense of their own destiny. [...] To discover the writings of these men is to discover, as it were, the birth of new principles in the human sciences.[15]

Missionary writers used their connections with the world of science in their evangelical rhetoric. In numerous mission fields, they wrote of their use of scientific explanations of natural phenomena to break down "heathen superstitions" about such issues. In providing scientific explanations for natural phenomena, the missionaries believed that they "proved" to the colonial convert the existence of the one true Christian God. This evangelical alliance

---

11 Williams, *Narrative of Missionary Enterprises*, 20–24.

12 *Narrative of Missionary Enterprises*, 23.

13 *Narrative of Missionary Enterprises*, 27–37.

14 Mary Louise Pratt, *Imperial Eyes: Travel Writing and Transculturation* (London & New York: Routledge, 1992).

15 Niel Gunson, Preface to *Australian Reminiscences and Papers of L.E. Threlkeld, Missionary to the Aborigines, 1824–1859*, by L.E. Threlkeld, ed., preface & intro. Niel Gunson (Australian Aboriginal Studies 40, Ethnohistory Series 2; Canberra: AIAS, 1974): v.

of rational modernity and religious faith is symptomatic of the doubled address of missionary discourse and textuality.

Missionary transformations of domestic spaces and arrangements were also seen as crucial indicators of Christian success. The sermon by the Rev. James Sherman already cited noted the importance of "architecture" in the christianization of the region and indeed the construction of Western-style homes was a crucial project of the LMS missionaries. The 1820 *Report* of the LMS attested that the "pernicious custom of herding together" had been solved when,

> Many of the natives, with their wives and children, are already living separately in neat habitations of their own; others have been induced to engage in preparing dwellings of a similar description; and it was expected, that in a short time there would be erected not less than twenty good houses, which would contribute to the respectable appearance of the village, as well as to the convenience, health, and morals of its inhabitants. (LMS *Report*, 15)

Missionary narratives continually noted the housing and town planning in their new mission station villages. Narratives of missionaries on tour were particularly explicit in their use of these issues as measures of "civilization and christianization." The LMS's 1824 *Report*, for example, noted that

> civilization continues to advance; more land has been brought into a state of cultivation. The erection of improved dwelling-houses proceeds; and during the last year a very superior house has been built for the Mission. In the manners and habits of the people there is a visible improvement. (LMS *Report*, 18)

The Rev. Barff reported on his 1830 visit to Maiaoiti that "a larger number of neatly plastered native houses had been erected. Each house was surrounded by a garden, well stocked with potatoes and other vegetables, and [...] the whole settlement presented the appearance of a neat village, in a state of great prosperity. Order, harmony, and industry prevailed" (LMS *Report* 1832, 13).

The LMS Deputation led by the Rev. Daniel Tyerman and George Bennet around the South Pacific mission stations of the LMS in the 1820s repeatedly remarked in their published *Journal* upon the physical state of these habitations. On approaching Rurutu, they "'were greatly surprised to see several neat-looking white houses at the head of the bay. From this we concluded that the Gospel had reached its shores'."[16] As they travelled, they

[16] Quoted in Williams, *Narrative of Missionary Enterprises*, 50.

approvingly noted the "new style, wattled and plastered" houses and public buildings,[17] and were explicit about why, in missionary eyes, housing styles were so important to christianization:

> The provident and well-regulated modes of living, introduced with the gospel, have proved favourable to improvement in every way, and perhaps in none more than in their domestic economy, from which, *decency, good order, and comfort have been expelled the grossness, confusion, and filthiness of what might be called promiscuous intercourse* – when men, women, and children, inmates and strangers, ate, drank, and lodged, in one long, narrow apartment, of which the whole structure consisted.[18]

The Rev. Aaron Buzacott on Rarotonga likewise stated:

> It took a considerable time to induce the natives to build separate houses, and *it was not until the gospel had quickened their moral natures that they were able to understand and to long for the comforts of a home.* In course of time the missionaries succeeded in overcoming all obstacles. The wretched huts were removed, and comfortable detached cottages for each family were erected on each side of the highroad. The new cottages were built of wood, wattled and plastered with lime obtained from the coral. Of course each dwelling was divided into several apartments, and adorned with venetian windows.[19]

It is evident from these missionary narratives that attempts to revolutionize Polynesian housing styles had multiple incentives. These included the missionary's eurocentric belief that British housing styles were naturally superior and also innately "moral" structures; that is, that their architecture somehow embodied Christian morality in its structuring of nuclear-family-based space and interpersonal relations. Missionaries in Polynesia genuinely believed that traditional housing styles, which were ideally suited to the climate and lifestyle of the area,[20] enabled and encouraged "heathen im-

---

[17] Rev. Daniel Tyerman & George Bennet, Esq., *Journal of Voyages and Travels by the Rev. Daniel Tyerman and George Bennet, Esq. Deputed from the London Missionary Society, to Visit their Various Stations in the South Sea Islands, China, India, &c., Between the Years 1821 and 1829*, comp. James Montgomery, 2 vols. (London: Westley & Davis, 1831), vol 1: 317.

[18] Tyerman & Bennet, *Journal of Voyages and Travels*, vol. 1: 164 (emphasis added).

[19] Quoted in Sir Peter Buck (Te Rangi Hiroa), *Mangaia and the Mission*, ed. & intro. Rod Dixon & Teaea Parima (Suva, Fiji: Institute of Pacific Studies, U of the South Pacific, in association with B.P. Bishop Museum, 1993): 42 (my emphasis).

[20] It is ironic that when the missionaries endured the extreme weather of the islands, such as the cyclones / hurricanes and monsoonal rains, they often had to abandon their

morality." Buck notes that Buzacott's reference to morality and housing embodied his belief that community houses were conducive to immorality, and argues that "it may have been so in Western communities, but it certainly was not in Polynesia. There was less chance of immorality in large single roomed houses than smaller house [sic] divided into rooms."[21]

Missionaries appeared to find Polynesian houses and their intimations of alternative life-styles and moral values highly unsettling. When left at Pangopango, the Rev. Murray wrote:

> Native houses have neither locks nor bars. In truth, we felt ourselves very insecure; but our apprehensions arose, not so much from our being in the power and at the mercy of the heathen, as from the character of the men [...] of our colour, and speaking our own tongue, whom we found upon the island.[22]

Other missionaries projected this insecurity onto their own social requirements for housing – John Williams rebuilt many of the houses at Rarotonga, stating: "It was my determination, when I originally left England, to have as respectable a dwelling as I could erect, for the Missionary does not go to barbarize himself, but to civilize the heathen. He ought not, therefore, to sink down to their standard, but to elevate them to his."[23]

The missionaries' own residences provided another example of the complex negotiations of space and place which occurred in Polynesian mission communities. As Figure 2 demonstrates, missionaries did not fail to provide themselves with the comforts of home. Indeed, in housing and in many other aspects of missionary life in the colonial field, LMS missionaries were able to enjoy considerably higher standards of living and social status than would have been available to them back in England. The caption to this illustration noted that the king's residence was modelled after the Rev. Buzacott's own house – it seems that in the colonial field, missionaries lived like

---

Western-style homes for the native-style buildings which had been built with the local conditions in mind; see the Rev. A.W. Murray, *Forty Years' Mission Work in Polynesia and New Guinea, from 1835–1875* (London: Nisbet, 1876): 147, 180.

21 Sir Peter Buck (Te Rangi Hiroa), *Mangaia and the Mission*, 42.

22 Murray, *Forty Years' Mission Work in Polynesia and New Guinea*, 29. See Vanessa Smith's excellent *Literary Culture and the Pacific: Nineteenth-Century Textual Encounters* (Cambridge Studies in Nineteenth-Century Literature and Culture; Cambridge: Cambridge UP, 1998) for a detailed examination of the troubled relations between European missionaries and beachcombers.

23 Williams, *Narrative of Missionary Enterprises*, 477.

kings, often before the kings themselves. It is also important to note the implications of these reorganizations of Polynesian space: the construction of a (fenced) missionary compound allowed the promenading and performance of European gentility, clearly separate from the Polynesian environment. In the lower picture, Polynesians are evident within the compound – as labourers. As is often the case, there is a considerable slippage between missionary rhetoric of equality before the eyes of god and the practice of cross-cultural relations in colonial environments. The sheer theatricality of this environmental reform is also significant in its foregrounding of the level of cultural performance in which the missionaries were engaged:

> It was a feature of Aoe's intrusions that they made small plays on the beaches. They came in islands of their own ships or they made islands of their own in mission stations and forts, or, if they were beachcombers, they let the beaches between their new islands and their old islands run down the middle of their lives. Confronted by what was different, exotic and to them bizarre, as well as bewildered that their own "natural" world was now unnatural and all their obvious symbols were meaningless, they played out their own cultural systems in caricatured charades.[24]

Such missionary attempts at transformation were significant not only because they introduced the kind of environmental and cultural change which is now causing real problems for island cultures in a climate of global capitalism and environmental degradation, but because the ways in which these missionaries made their spaces in these island cultures tell us a lot about the impetus and intentions of missionaries. Nineteenth-century missionaries carved out spaces on these beaches in particular ways – very different, for example, from the ways that European beachcombers did – and in so doing they attempted to institute particular kinds of social relations. This project of cultural as well as environmental reform had a significant impact on the local communities. It engendered a certain set of social hierarchies, of (racialized) labour relations, of gendered work, and of relationships between people and their natural environment. Of course, the missionaries' attempts to transform cultures and environments were rarely realized in quite the manner they intended:

> There were few things that came across the beach that seemed to change a value or an institution. The process of change was much more indirect. Aoe artefacts changed Enata institutions when their distribution affected established

24 Dening, *Islands and Beaches*, 18–19.

> relationships among Enata or when, to acquire the artefacts, Enata had to change their behaviour, or when the accumulation of these artefacts eliminated the necessity of some local industry with its associated networks of exchange and relationships. [...] One cannot point to a cultural importation or a mimicry that triggered a collapse of Enata culture. Rather, individuals made personal adaptations which widened their options in behaviour and lessened the control of what was culturally given to them over their actions. In this cultural agnosticism the force of sanctions weakened, the need for established roles decayed. Cultural change became the infinite number of choices exercised concerning an environment that was changed both by the intrusions and by the choices themselves.[25]

Missionaries naturally reported what they wanted to see, but by the 1830s John Williams, when visiting one of the settlements near the station at Amoa, praised the village in detail. His loftily approving description highlights the complex ways in which environmental reform – in its many facets – was believed by the missionaries to effect moral, social, and even racial reform:

> we passed through a settlement called Safatulafai, which is one of the most beautiful in the group, and which astonished and delighted me. We could more easily have imagined ourselves in an English part, than in a heathen village. [...] This settlement was kept in excellent order, and had an air of respectability which could not have been looked for among a people, in other respects, so barbarous.[26]

WORKS CITED

Buck, Sir Peter (Te Rangi Hiroa). *Mangaia and the Mission*, ed. & intro. Rod Dixon & Teaea Parima (Suva, Fiji: Institute of Pacific Studies, U of the South Pacific, in association with B.P. Bishop Museum, 1993).

Dening, Greg. *Islands and Beaches: Discourse on a Silent Land; Marquesas, 1774–1880* (Chicago: Dorsey, 1980).

Ellis, William. *Polynesian Researches, During a Residence of Nearly Six Years in the South Sea Islands; Including Descriptions of the Natural History and Scenery of the Islands – with Remarks on the History, Mythology, Traditions, Government, Arts, Manners, and Customs of the Inhabitants*, 2 vols. (London: Fisher & Jackson, 1829).

---

[25] Dening, *Islands and Beaches*, 125.

[26] Williams, *Narrative of Missionary Enterprises*, 445–46.

Gill, Rev. William Wyatt. *Life in the Southern Isles: Or, Scenes and Incidents in the South Pacifc and New Guinea* (London: Religious Tract Society, 1876).

Grove, Richard. *Green Imperialism: Colonial Expansion, Tropical Island Edens, and the Origins of Environmentalism, 1600–1860* (Studies in Environment and History; Cambridge: Cambridge UP, 1995).

Gunson, Niel. Preface to *Australian Reminiscences and Papers of L.E. Threlkeld, Missionary to the Aborigines, 1824–1859*, by L.E. Threlkeld, ed., preface & intro. Niel Gunson (Australian Aboriginal Studies 40, Ethnohistory Series 2; Canberra: AIAS, 1974): v–vii.

Kent, Graeme. *Company of Heaven: Early Missionaries in the South Seas* (Wellington: Reed, 1972).

LMS. *Reports of the London Missionary Society*, 1796–1870. School of Oriental and African Studies Collection, University of London, UK.

Lovett, Richard. *The History of the London Missionary Society, 1795–1895*, 2 vols. (London: Frowde, 1899).

Murray, Rev. A.W. *Forty Years' Mission Work in Polynesia and New Guinea, from 1835–1875* (London: Nisbet, 1876).

Neill, Stephen. *A History of Christian Missions* (Penguin History of the Church; Harmondsworth: Penguin, 1964).

Pratt, Mary Louise. *Imperial Eyes: Travel Writing and Transculturation* (London & New York: Routledge, 1992).

Sharrad, Paul. "Putting Down Roots: Colony as Plantation," *SPAN* 46 (April 1998): 72–86.

Sherman, Rev. James. *The Remembrance of Christ's Love a Stimulus to Missionary Exertion: A Sermon, Preached before the London Missionary Society, at Spa Field's Chapel, on Thursday, May 13th, 1830* (London: Fisher & Jackson, and Nisbet, n.d.).

Smith, Vanessa. *Literary Culture and the Pacific: Nineteenth-Century Textual Encounters* (Cambridge Studies in Nineteenth-Century Literature and Culture; Cambridge: Cambridge UP, 1998).

Thomas, Nicholas. *Entangled Objects: Exchange, Material Culture, and Colonialism in the Pacific* (Cambridge MA: Harvard UP, 1991).

Tyerman, Rev. Daniel, & George Bennet, Esq. *Journal of Voyages and Travels by the Rev. Daniel Tyerman and George Bennet, Esq. Deputed from the London Missionary Society, to Visit their Various Stations in the South Sea Islands, China, India, &c., Between the Years 1821 and 1829*, comp. James Montgomery, 2 vols. (London: Westley & Davis, 1831).

Williams, John. *A Narrative of Missionary Enterprises in the South Sea Islands: With Remarks upon the Natural History of the Islands, Origin, Languages, Traditions, and Usages of the Inhabitants* (1837; London: Snow, 1838).

A VILLAGE IN PUKAPUKA, UNDER HEATHENISM

THE SAME VILLAGE, UNDER CHRISTIANITY.

Figure 1

Pukapuka under 'heathenism' and under Christianity, from William Wyatt Gill, *Life in the Southern Isles: Or, Scenes and Incidents in the South Pacific and New Guinea* (London: Religious Tract Society, 1876): pages 18–19. Courtesy of Fryer Library, University of Queensland Library

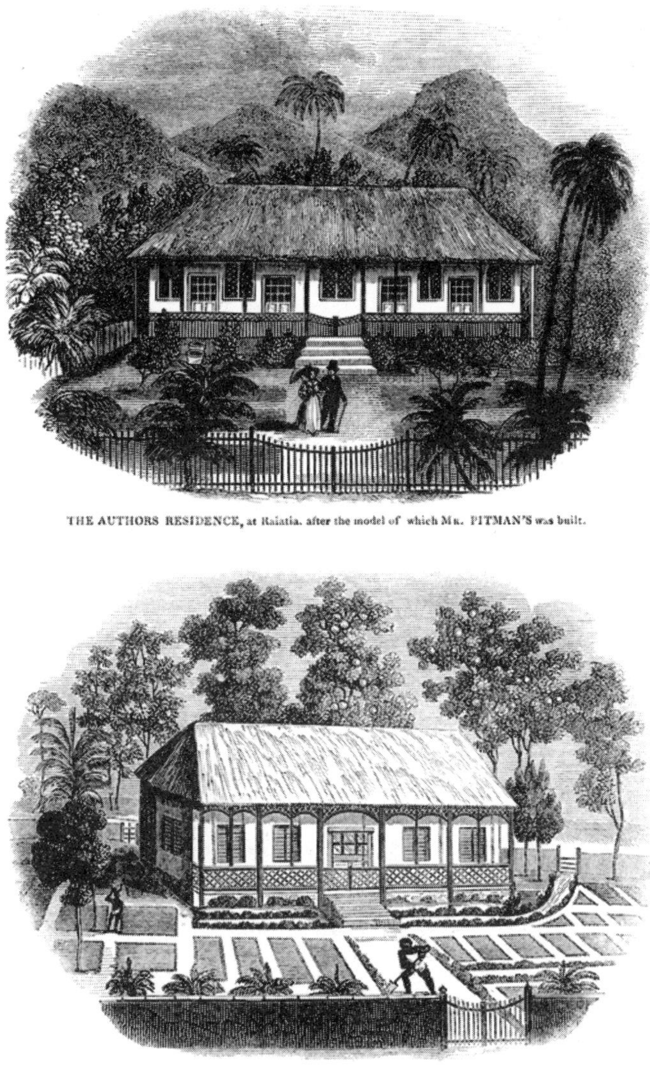

Figure 2

Missionary houses, from John Williams, *A Narrative of Missionary Enterprises in the South Sea Islands: With Remarks Upon the Natural History of the Islands, Origin, Languages, Traditions, and Usages of the Inhabitants* (London: J. Snow, 1837): illustration facing page 475. Courtesy of Fryer Library, University of Queensland Library

Five Emus to the King of Siam

Acclimatization and Colonialism

CHRIS TIFFIN

AT ITS SEPTEMBER 1865 MEETING, the Queensland Acclimatisation Society reported on its activities for the previous month which included donations received and specimens despatched. Among the disbursements were two kangaroos to the Royal Zoological Society in London, two scrub turkeys to New Caledonia, three red deer to a local landholder to try to breed from them, fish from the Mary River sent to Tasmania, and "five emus to the King of Siam."[1] Such a flurry of activity testifies to the enthusiasm with which the project of acclimatization was pursued, and raises the question – what motivated prominent private citizens to embrace such an improbable task so enthusiastically? This discussion is concerned principally with the ideology of Acclimatization: how the acclimatizers saw their mission, and how it related to colonialism.

"Acclimatization" strictly means the adaptation of individual specimens to a new environment, but the loose meaning of the word in the 1860s was the introduction of plants, animals, birds and insects into areas to which they are not endemic. As Edward Wilson said, "What we seek to do is rather to *distribute* the good things of the earth than necessarily to

[1] Archive of the Queensland Acclimatisation Society. John Oxley Library, Brisbane. OM66-24/2, Scrapbook 2.

acclimatise them."[2] The process is so well-established in human history that it is sometimes hard to remember that plants and animals that have become almost emblematic of a particular country, such as the potato in Ireland or the sheep in New Zealand, are actually imports. By the mid-nineteenth century some of the settler colonies of the British Empire, notably Canada and Australia, had already established whole economic systems on the basis of agricultural and pastoral imports. When the nascent Queensland Acclimatisation Society petitioned the colonial Legislative Council for a grant, it was able to argue validly that "the chief producing interest in the colony owes its origin to the importation of animals not indigenous to Australia, and that almost all future products, such as sugar, coffee, tobacco, cotton &c., whether for home consumption or exportation will have similar originations."[3] Food crop imports, in fact, were sometimes startlingly experimental. By 1863 the Horticultural Society of Victoria (founded in 1849) had 333 different species of apple under cultivation.[4]

Although by the 1860s acclimatization was an ancient practice, it was then topical as it had been recently rebadged and refocused. The new movement was born in France in 1854 when Isidore Geoffroy Saint-Hilaire founded a society to promote botanical and zoological exchange with Algeria, and a few years later established the Jardin d'Acclimatation on the outskirts of Paris.[5] The British were quick to follow this lead, although, as Christopher Lever points out, the Zoological Society of London, founded in 1826, had included the fundamental idea of acclimatization in its Prospectus.[6] The six-point prospectus of the British Acclima-

---

[2] Edward Wilson, *Acclimatisation: Read Before the Royal Colonial Institute* (London: Unwin Bros., 1875): 2.

[3] Archive of the Queensland Acclimatisation Society. John Oxley Library, Brisbane. OM66-24/1, Scrapbook 1.

[4] Horticultural Society of Victoria, *Catalogue of Fruit Trees, Experimental Gardens, Survey Paddock* (Richmond, Melbourne: Wilson & McKinnon, 1863).

[5] For a good introduction to the Acclimatization movement, see Christopher Lever, *They Dined on Eland: The Story of the Acclimatisation Societies* (London: Quiller, 1992); R.M. McDowall, *Gamekeepers for the Nation: The Story of New Zealand's Acclimatisation Societies 1861--1990* (Christchurch: U of Canterbury P, 1994), and F.H. Jenkins, *The Noah's Ark Syndrome: One Hundred Years of Acclimatization and Zoo Development in Australia* (Perth: Zoological Gardens Board, W.A., 1977).

[6] Lever, *They Dined on Eland*, 24.

tisation Society, founded in 1860, gives a representative view of how the Acclimatization Movement saw its role and function:

> The purposes of the Society are:
>
> i) the introduction, acclimatisation, and domestication of all innoxious animals, birds, fishes, insects, and vegetables whether useful or ornamental
>
> ii) the perfection, propagation and hybridisation of races newly introduced or already domesticated
>
> iii) the spread of indigenous animals etc. from parts of the United Kingdom where they are already known, to other localities where they are not known
>
> iv) the procuration, whether by purchase, gift, or exchange, of animals etc from British Colonies and foreign countries
>
> v) the transmission of animals etc from England to her colonies and foreign parts, in exchange for others sent thence to the Society
>
> vi) the holding of periodical meetings, and the publication of reports and transactions for the purpose of spreading knowledge of acclimatisation, and inquiry into the causes of success or failure.[7]

The Movement, then, proposed three activities: the redistribution of plants and animals; the selective breeding of plants and animals; and the propagation of knowledge about these processes.

The Acclimatisation Societies occupied a peculiar niche in colonial administration. Although essentially a private initiative, they were founded by individuals keen to enlist both government and commercial support. In this they were fairly successful. The British Government and colonial administrations in Australia all made cash grants to the Societies and the Royal Navy was directed to assist with the transport of plant and animal specimens. Societies were permitted to hold meetings in the Legislative Chambers and Municipal offices of colonies and cities; they negotiated directly with other colonial and foreign governments; they brokered concessional deals with shipping firms, and secured publicity and sponsorship from colonial newspapers.[8] In their heyday, Acclimatisation Societies were

---

[7] Society for the Acclimatisation of Animals, Birds, Fishes, Insects and Vegetables Within the United Kingdom, *First Annual Report* (London: The Society, 1861): 4.

[8] The Acclimatisation Societies were greatly assisted by the active patronage of colonial governors such as Sir Henry Barkly and Sir George Bowen. The Queensland Society was formed at Bowen's instigation, while Barkly was the inaugural President of

seen as working *pro bono publico* and consequently were given the sort of acknowledgement accorded more recently to Live Aid or World Cleanup Day. It was not atypical that the Admiralty Survey Vessel was coopted by the Queensland Society to plant pandanus and mangoes on the islands off the Queensland coast.[9]

Acclimatization was largely concentrated in the French and English empires, and, as Lloyd Osborne wrote, "No British territory boasted more voluntary acclimatization activity than Australia."[10] By 1862 there were individual societies in New South Wales, Victoria, Queensland and South Australia. Their endeavours were situated among, and interacted with, those of zoological societies and botanic gardens, those of philosophical and natural history societies, and those of pastoral and agricultural associations. That is to say, acclimatization activity overlapped in aims and membership with groups interested in collecting and displaying plants and animals, with groups seeking and disseminating knowledge about plants and animals, and with groups raising plants and animals commercially. The distinguishing activity of the Acclimatisation Societies, however, was the introduction and propagation of plants and animals for release into the environment. The intent was not just to import for study and display, but rather to enrich the environment with the new species.

As the Societies were essentially private concerns, they sometimes encountered charges of amateurism and even professional jealousy. While the Victorian Society was headed by two of the colony's most important scientists, the Director of the Botanic Gardens, Dr Ferdinand Mueller, and the palaeontologist with the Government Survey, Professor Frederick McCoy, the Queensland Society seems to have had a less happy relationship with that colony's scientists. Initially the Botanic Gardens provided space for the Society's activities until it acquired its own land at Bowen Hills, but within a few years tensions appeared between the Society and the Director

---

the Victorian Society, and continued his Acclimatization activities when he became Governor of Mauritius.

[9] Queensland Acclimatisation Society Council Meeting 19 November 1875. John Oxley Library OM66-24/3. Scrapbook 3.

[10] Michael A. Osborne, "A Collaborative Dimension of the European Empires: Australian and French Acclimatization Societies and Intercolonial Scientific Co-operation," in *International Science and National Scientific Identity: Australia Between Britain and America*, ed. R.W. Home & Sally Gregory Kohlstedt (Dordrecht: Kluwer Academic 1991): 105.

of the Botanic Gardens, Walter Hill. In one of the Society's cutting books, a letter by Hill to the Brisbane *Courier* of 12 August 1876 concerning the importation of jute has been endorsed by the President, L.A. Bernays, "Mr Hill stealing our project." The following year further tension developed when the Society learned that one Count Giuseppe Francheschi had imported a number of grapevines from southern Europe. Fearing an outbreak of *phylloxera vastarex* which had been ravaging European vines since 1865 but was unknown in Queensland, the Society advised the government of the danger and called on it to persuade the Count to destroy his vines.[11] This intervention by a group of amateurs was apparently too much for Hill who sided with Francheschi, saying that he (Hill) had examined the vines on arrival and that he saw no sign of the disease. Moreover, he noted that the Queensland Museum (with whom the Acclimatisation Society was collaborating on fish breeding) had recently imported and distributed grape seed from Germany, which was far more likely to spread disease since it would not become manifest until the plants had struck. In a clever but provocative move to assert his dominance, Hill endorsed the Society's concern about introduced plant diseases and suggested that these could be avoided if all future importations of plant material by individuals or institutions were subject to inspection by him.

My interest, though, is less in the official history and unofficial politics of the Acclimatisation movement than in its mind-set, or range of mind-sets. How did it justify its aims? With what sort of rhetorical gestures did it proclaim them? How did it see the people–land nexus? What was its role in the process of colonialism?

Acclimatisation Societies sought four principal outcomes:

i) rendering the environment richer and more attractive:
ii) providing quarry for "sport" (hunting and fishing):
iii) increasing the variety of food sources and other useful bioproducts:
iv) increasing knowledge of the natural world.

The first of these aims, rendering the environment richer and more attractive is the one which is best known and most often satirized. Certainly, some of the enthusiasms do seem just a little risible. For example, Edward Wilson, proprietor of the Melbourne *Argus* and a leading acclimatizer, proposed the introduction of monkeys into the Victorian bush "for the amusement of the wayfarer whom their gambols would delight as he lay

[11] Meeting of the Queensland Acclimatisation Society, 1 February 1877.

under some gum tree in the forest on a sultry day." [12] This vignette may have derived from Wilson's recent South American experience, but its discursive resonance seems classical, perhaps via landscape painting or bucolic poetry. Wilson invokes a sense of physical and temporal disengagement in which the landscape provides an innocuous *divertissement* rather than confronting the traveller with the difficulty, tediousness or danger that the real journey might involve.[13]

In a similar vein, the cashmere goat was recommended by a member of the NSW Acclimatisation Society as being "a fine looking animal [which] would be very ornamental in a park, or a ruin, or the side of a rock, or in a churchyard."[14] Again there is something painterly about the prospect. The goat is not recommended as a way of retrieving a real lost European landscape but rather to invoke a fantasy of classical landscape peopled by a heraldic animal. The improvement foreseen is an evocation of a pseudo-cultural memory constituted through tokens of order, age, and venerable human dissolution.

The justification for enriching the landscape is unashamedly biblical. In the view of Edward Wilson, the Creator's injunction to "be fruitful and multiply and replenish the earth, and subdue it," has been sadly neglected. It turns out, moreover, that in the Bible, God had been speaking directly to Imperial Britain:

> I consider that it is a waste of opportunity to have done so little, and that, having paved the way by opening up fresh countries by conquest and colonisation, the grand scheme of distribution should be elaborated scientifically, systematically, and exhaustively. I hold that we should never rest until every country on earth is duly furnished with every good thing which that country is capable of maintaining.
>
> And here I wish to say explicitly that in the term "good" I refer to things not simply useful in a practical and economic, or mercantile sense, but good in the sense of adding in any way to the legitimate enjoyment of mankind.[15]

---

[12] Wilson is being approvingly paraphrased by Sir Henry Barkly. Quoted by Jenkins, *The Noah's Ark Syndrome*, 14–15.

[13] Wilson subsequently introduced monkeys on to an island in a lake on his English property with evident success. "I really do not know when I have seen anything more enjoyable than the darting about over the upper branches of the trees of these playful and beautiful creatures"; Wilson, *Acclimatisation*, 16.

[14] Meeting of the NSW Acclimatisation Society, 13 March 1865.

[15] Wilson, *Acclimatisation*, 4.

According to some acclimatizers, adding "to the legitimate enjoyment of mankind" concomitantly improved its morals. In a deeply Wordsworthian flight, Frederick McCoy, Professor of Natural Sciences at the University of Melbourne, argued for the moral impact of European songbirds, through which

> the plains, the bush, and the forest would have [...] their present savage silence, or worse, enlivened by their varied touching, joyous strains of Heaven-taught melody, which our earliest records show, have always done good to man, which, in all times have been recognized, among all varieties of nation or taste, as sweetening the poor man's labours, inspiring the poet with happiest thoughts, and softening and turning from evil even the veriest brute that ever made himself drunk or plotted against his neighbour. [16]

Two aspects of this line of thought are interesting. In the first place, McCoy asserts that the European songbird is universally appealing – all nations, and all tastes, in all times have felt its beneficent effects. Moreover, the reason for this universality of aesthetic appeal and moral efficacy is that it is "Heaven-taught," because natural. What the formulation cannot then explain is why other natural (read: heaven-taught) birdcalls, such as those of the crow or the sulphur-crested cockatoo, are not likewise corrections against bibulous excess or neighbourhood malice.

Claims of the moral effects of acclimatization crop up in the strangest places. G.W. Francis, Director of the South Australian Botanic Gardens, vigorously supported the importation of asses to breed mules to replace bullocks, for the curious reason that he thought the slow speed of the bullock-wagons induced a psychic rhythm in the boys who drove them which would prevent them, when they grew up, of ever moving faster than two miles an hour.[17]

The biblical framework within which acclimatization thought was developed ensured that the comfort and safety of humans remained prominent. Consequently, acclimatization could call for the eradication of species which were seen as detrimental to human comfort and safety. As Edward Wilson put it, "We ought, I think, to assert our right to destroy some things for the purposes of smoothing the path of more valuable things." Thus, in the rhetoric of acclimatization, birds which attack crops are "criminal," and Wilson is shocked that anyone would oppose the elimi-

---

[16] Quoted by Jenkins, *The Noah's Ark Syndrome*, 15.

[17] Jenkins, *The Noah's Ark Syndrome*, 46--47.

nation of the Asian tiger, which allegedly killed ten thousand Indians a year.[18] The other side of this particular debate shows how ethnocentric the discussion could be, and also how imbricated it was in the imperial process. Those who opposed the wholesale slaughter of the tigers did so on the grounds that such a campaign would eliminate one of the principal recreations of the British in India, tiger-hunting, without which life would be so dull there that it would be impossible to recruit staff for the Indian Civil service. The sort of class issues which were raised in the moral improvement of the "drunken brute" enter this issue with a racial inflection. The killing of Indians by tigers is a necessary condition for allowing the killing of tigers by Englishmen.

Wilson himself was opposed to hunting,[19] but for many Acclimatization thinkers, the provision of hunting quarry was a major, if not the major, goal of the movement. The seventy-year-long rabbit plague in Australia was the most infamous example of a hunting quarry importation, although actually the Acclimatisation Societies were not responsible for this mishap. They did, however, import hares, gamebirds, and a number of species of deer, including one small herd donated by Queen Victoria.

Harriet Ritvo has argued that big-game hunting offers an emblem of colonial domination, to the point that "ultimately, the hunter emerged as both the ideal and the definitive type of empire builder."[20] Australian fauna proved rather an unsatisfactory emblem, which the Acclimatisation Societies were keen to improve. In the meantime, however, most visitors to Australia reported the shooting as unsatisfactory. Richard Twopenny complained that there was no fox-hunting, the country was "monotonous and wants very good riding," "the rabbits are generally too plentiful," and "the kangaroos and wallaby are generally too tame."[21] Another visitor, Douglas Sladen, nephew of the Premier of Victoria, explicitly mapped his hunting experiences in Australia as the unsatisfactory fulfilment of expectations aroused by imperial reading as a child:

> Wandering about the big paddocks gun in hand [...] had something of the excitement of the books about the American backwoods which I had read in

---

[18] Wilson, *Acclimatisation*, 20.

[19] "I cannot understand the pleasure of inflicting death"; Wilson, *Acclimatisation*, 21.

[20] Harriet Ritvo, *The Animal Estate: The English and Other Creatures in the Victorian Age* (Cambridge MA: Harvard UP, 1987): 254.

[21] Richard Twopenny, *Town Life in Australia* (London: Stock, 1883): 208--209.

> my childhood. It is true that I would rather have shot grizzly bears than the native bears of Australia, mere sloths, and lions and tigers than kangaroos, but a big "forester" is not to be sneezed at.[22]

The third aim of the Acclimatisation Societies, that of increasing the variety of food resources and other "useful" bio-products was the most important one. Underpinned by the belief that any eco-system was infinitely elastic and receptive to new species, it had little sense that environments could be fragile and easily depleted. Instead, it prided itself on the democratic and egalitarian tendency of its cornucopian philosophy. The Queensland Society aspired to "the improvement of the food of the people, and the making common upon the table of the poor man, delicacies which are only now attainable by the rich, and others which are at present unattainable by either."[23]

The final object of the Societies was to understand better the plants and animals they were importing. This was partly functional – the more the Societies knew about their specimens, the better chance they had of raising them successfully. Consequently they sought advice from other societies and from Botanic Gardens and zoo curators, compiled scrap books of information, and maintained libraries of botanical and zoological books and journals. But they also saw environmental knowledge as necessary for effective colonization. The Queensland Society petitioned its government to increase the teaching of science in schools, wanting

> the next generation of colonists to know, amongst other things, more about the structure and physiology of plants, and the scientific principles involved in the constituent parts of soils, in rainfall and its influence, and in a number of other things which would make them better agriculturalists and acclimatisers.[24]

Historians of science point out that for all their advocacy of scientific principles, the Acclimatization societies were generally slow to take up Darwinian ideas of selection and species development.[25] Their usual position was that each species was perfected and that acclimatization did not involve adaptation so much as redistribution of the species to an environment similar to that in which they were endemic. Frederick McCoy, for example, elaborated a "theory of representative forms or species" to argue

---

[22] Douglas Sladen, *Twenty Years of My Life* (London: Constable, 1915):18.

[23] Meeting 13 May 1863.

[24] Meeting 9 September 1876.

[25] Osborne, "A Collaborative Dimension of the European Empires," 107.

that similar animals in comparable climatic enclaves had been created thus, and had not evolved from any common ancestor. However, their similarity did make them available for redistribution, and the real principle of successful acclimatization was to identify the animals which fell into each group of "representative forms" and place all members of each group into all locations occupied by any member of that group.

Acclimatization and colonization were too closely parallel, however, to remain distinct in the thought of the time, and when Darwinian ideas were deployed to explain and justify the elimination of the native races in Australia, New Zealand and North America, questions of genetic superiority and adaptation in human and animal worlds were reinvoked. Charles Dilke, one of the leading figures in the Greater Britain movement, published a book about his visit to Australia and New Zealand in 1867. He found the Australian Aborigines and the New Zealand Maori to be declining races inevitably to be overcome by the greater vitality and biological robustness of the English. In an argument that runs somewhere between social Darwinism and homoeopathy, Dilke claimed that several centuries of imperial adventure had strengthened the British organism through exposure and struggle so that the less seasoned Australian and New Zealand communities were no match for it.

> That which is true of our animal and vegetable productions is true also of our man. The English fly, grass and man, they and their progenitors before them, have had to fight for life against their fellows. The Englishman, bringing into his country from the parts to which he trades all manner of men, of grass seeds, and of insect germs, has filled his land with every kind of living thing to which his soil or climate will afford support. Both old inhabitants and interlopers have to maintain a struggle which at once crushes and starves out of life every weakly plant, man, or insect, and fortifies the race by continual bufferings. The plants of civilized man are generally those which will grow best in the greatest variety of soils and climates; but in any case, the English fauna and flora are peculiarly fitted to succeed at our Antipodes, because the climate of Great Britain and New Zealand are almost the same, and our men, flies and plants – the "pick" of the whole world – have not even to encounter the difficulties of acclimatization in their struggle against the weaker growths indigenous to the soil.[26]

---

[26] Charles Dilke, *Greater Britain*, ed. Geoffrey Blainey (North Ryde, N.S.W.: Methuen Haynes, 1985): 76.

For Dilke, acclimatization is not a new programme of cornucopian enrichment of the environment, but rather a longstanding component of the process of competitive trade and colonization. Its mandate derives not from the incomplete fiat of creation but rather from the homoeopathic vitalism of imperial expansion. Abandoning the rhetoric of pleasure, ethics, and even utility for that of contest and triumphalism, he finds acclimatization a constant, if unseen force in the teleology of empire.

With the benefit of ecological hindsight, recent historians of the Acclimatization movement dismiss it as a silly, misguided enterprise, saved from being seriously destructive only by its own incompetence. As Eric Rolls puts it, "The history of the Societies is more a history of damage that might have been done; for there never was a body of eminent men so foolishly, so vigorously, and so disastrously wrong."[27]

But the strange phenomenon of the Acclimatization movement does provide a unique window on the colonizing eye of imperial England: how it looked at the lands it conquered and occupied, and how it sought to develop and redistribute the biological riches of the planet. Although limited by its amateurism and its inability to envisage ecology in any comprehensive way, the Acclimatization movement canvassed ideas which included enriching food stocks for the nation, biological control of pests, animal and plant development, the ethical effect on humans of the natural world, cultural prejudices in nutrition, and the preservation of endangered species. These are issues which are neither foolish nor trivial. Moreover, the movement effectively marshalled public and private support in the colonies and established a truly international network for biological barter. It offered a benevolent vision of an improved animal and vegetable world, and in doing so paralleled and reinforced the contradictions of the colonizing project itself. The Acclimatization movement was a significant, if angled, reflection of nineteenth-century European expansion and empire.

---

[27] Eric Rolls, *They All Ran Wild: The Story of Pests on the Land in Australia* (London: Angus & Robertson, 1969): 210.

WORKS CITED

Dilke, Charles. *Greater Britain*, ed. Geoffrey Blainey (North Ryde, N.S.W.: Methuen Haynes, 1985).

Horticultural Society of Victoria. *Catalogue of Fruit Trees, Experimental Gardens, Survey Paddock* (Richmond, Melbourne: Wilson & McKinnon, 1863).

Jenkins, F.H. *The Noah's Ark Syndrome: One Hundred Years of Acclimatization and Zoo Development in Australia* (Perth: Zoological Gardens Board, W.A., 1977).

Lever, Christopher *They Dined on Eland: The Story of the Acclimatisation Societies* (London: Quiller, 1992).

McDowall, R.M. *Gamekeepers for the Nation: The Story of New Zealand's Acclimatisation Societies 1861-1990* (Christchurch: U of Canterbury P, 1994).

Osborne, Michael A. "A Collaborative Dimension of the European Empires: Australian and French Acclimatization Societies and Intercolonial Scientific Co-operation," in *International Science and National Scientific Identity: Australia Between Britain and America*, ed. R.W. Home & Sally Gregory Kohlstedt (Dordrecht: Kluwer Academic 1991): 97–119.

Ritvo, Harriet. *The Animal Estate: The English and Other Creatures in the Victorian Age* (Cambridge MA: Harvard UP, 1987).

Rolls, Eric. *They All Ran Wild: The Story of Pests on the Land in Australia* (London: Angus & Robertson, 1969).

Sladen, Douglas. *Twenty Years of My Life* (London: Constable, 1915).

Society for the Acclimatisation of Animals, Birds, Fishes, Insects and Vegetables Within the United Kingdom, *First Annual Report* (London: The Society, 1861).

Twopenny, Richard. *Town Life in Australia* (London: Stock, 1883).

Wilson, Edward. *Acclimatisation: Read Before the Royal Colonial Institute* (London: Unwin Bros., 1875).

"Back to the World"

Reading Ecocriticism in a Postcolonial Context

SUSIE O'BRIEN

NINETEEN-SEVENTY-EIGHT WAS A PRODUCTIVE YEAR in English literary studies. It was the year of the publication of Edward Said's *Orientalism: Western Conceptions of the Orient*, a text which marked the definitive emergence into international (in particular US) currency of arguments about the connection between knowledge – even the so-called politically innocent knowledges of English literature and English literary scholarship – and colonial structures of domination. Postcolonialism was not the only body of theory spawned from literary critics' struggle to integrate their work as scholars with the radical cultural and political movements launched over the preceding decade. 1978 also marked the emergence from English literary studies of another stream of critical thought, this one clearly originating in the USA. In an essay which, unlike *Orientalism*, received little attention when first published, William Rueckert argued for the need to understand literature and literary criticism in the context of ecology. Like *Orientalism*, Rueckert's essay was not entirely original in its formulation of a particular approach to engaging the worldliness of English literature;[1] it did,

[1] William Rueckert, "Literature and Ecology: An Experiment in Ecocriticism," in *The Ecocriticism Reader*, ed. Cheryl Glotfelty & Harold Fromm (Athens: U of Georgia P, 1996): 105–23. See, for example, Joseph Meeker, whose *The Comedy of Survival: Studies in Literary Ecology* (New York: Charles Scribner's Sons, 1972) makes a convincing argument for the value and viability of reading generic codes of literature within an ecological framework.

however, offer a word which, twenty years later, has been taken up as the badge of a new critical movement – ecocriticism.

In this analysis, I want to look at ecocriticism as one of the more obscure intellectual fruits of the long and violent engagement of colonialism with the physical environment. By 'obscure' I want to suggest not just that ecocriticism is not well-known, although this is perhaps still the case outside the USA and the UK,[2] but also that its colonial lineage is not immediately obvious, even though, as this essay will argue, the concerns and limitations of ecocriticism are shaped by its affiliation with the postcolonial science of ecology. A small but increasingly well-recognized field, ecocriticism is worth investigating as a critical movement that has the potential to contribute to understanding of the ways in which cultures and environments shape each other – an understanding that becomes particularly urgent in the current period of environmental crisis. It also has a number of significant problems, which I want to focus on here as symptoms not so much of its inadequacy as a vehicle for environmentalist politics (though the burgeoning of ecocriticism has coincided with the global dismantling of environmental protections) as of its limitations as a critical discourse. These limitations, I suggest, turn out to be not unconnected to the problem of accelerating environmental degradation. I want to look at ecocriticism, finally, in the context of changes that have occurred in the critical climate (not to mention the global climate) since 1978 – changes which have also been registered in postcolonial criticism – and to suggest some productive linkages between the two.

As an ecologist who worked as a ranger in the National Park Service and also holds a PhD in comparative literature, Meeker was ideally and uniquely placed to draw out connections between ways of understanding nature and culture.

[2] Ecocriticism continues to be concentrated primarily in (and on) the USA, though ASLE (the Association for the Study of Literature and Environment), which emerged in 1992 as a branch of the Western Literature Association, has developed a significant membership in the UK, where it has begun to hold its annual conference in alternate years. The 1998 collection *Writing the Environment: Ecocriticism and Literature*, ed. Richard Kerridge & Neil Sammells (London: Zed, 1998), draws most of its contributors from the UK.

What is Ecocriticism?

Ecocriticism is defined in the introduction to *The Ecocriticism Reader* as "the study of the relationship between literature and the physical environment."[3] In his essay "Some Principles of Ecocriticism," included in the *Reader*, William Howarth offers a more detailed, if somewhat tongue-in-cheek, gloss on *ecocritic* (working from the derivation of 'eco' and 'critic'): "a person who judges the merits and faults of writings that depict the effects of culture upon nature, with a view toward celebrating nature, berating its despoilers, and reversing their harm through political action."[4] Howarth's definition, however ironically self-conscious, goes some way towards identifying the complex mixture of aims and methodologies which shape ecocriticism into a discourse at once political, aesthetic and scientific. In order more clearly to identify these strands, I want to return to Rueckert's pioneering essay on ecocriticism, an essay which, while it does not occupy the kind of seminal position in ecocriticism that *Orientalism* does in postcolonialism, does nevertheless set out the chief concerns which continue to inform contemporary ecocriticism. These concerns are embodied in the essay's three epigraphs:

> 1). "It is the business of those who direct the activities that will shape tomorrow's world to think beyond today's well being and provide for tomorrow."
> — Raymond Dasmann, *Planet in Peril*

The first epigraph speaks to what is perhaps the strongest impulse of ecocriticism, one which is, in the broad sense of the word, political. The fundamental premise grounding Rueckert's argument for ecocriticism is that the environment is in a state of crisis, largely of human making, and that urgent action is required if future disaster, encompassing humans and other species, is to be averted. This concern is reiterated by more recent ecocritics such as Lawrence Buell, who grounds (and legitimates) his study of Thoreau, nature writing and the formation of American culture in an entreaty: "I hope I do not need to spend many pages defending the reasonableness of the claim that [in former US Vice-President Al Gore's words] 'we must make the rescue of the environment the central organizing principle for civiliza-

---

[3] Cheryl Glotfelty, "Introduction: Literary Studies in an Age of Environmental Crisis," in Glotfelty & Fromm, ed. *The Ecocriticism Reader*, xviii.

[4] William Howarth, "Some Principles of Ecocriticism," in *The Ecocriticism Reader*, ed. Glotfelty & Fromm, 69.

tion'."[5] The quotations with which Rueckert and Buell choose to frame their arguments are striking in that their polemic is addressed, in a way unusual for literary criticism, quite specifically to global policy. For Buell, that the USA will be instrumental in formulating and implementing such a policy seems to go without saying. Rueckert's citation from Dasmann, read in a contemporary context, could be applied most accurately not to any particular nation's government, most of which seem to have lost or relinquished the capacity meaningfully to direct the activities that shape today or tomorrow's world, but rather to the corporate sphere.

Appeals to global policy-makers (whose world-shaping *authority*, it should be noted, is not questioned by Rueckert or Buell) are combined and sometimes conflated in much contemporary ecocriticism with the argument that environmental responsibility begins with the individual and, in particular, in this context, the individual academic. Thus Rueckert's call for policy initiatives is offset, with no apparent sense of contradiction, against the almost comically earnest injunction that "each individual has a responsibility for the entire biosphere and is required to engage in creative and cooperative activities."[6] Cheryl Glotfelty is more precise in her argument for academic responsibility, arguing that "as environmental problems compound, work as usual seems unconscionably frivolous," underlying her point with the now somewhat facile formula "if we're not part of the solution, we're part of the problem."[7] Glen Love offers strong support for Glotfelty's position, asking: "given the fact that most of us in the profession of English would be offended at not being considered environmentally conscious and ecologically aware, how are we to account for our general failure to apply any sense of this awareness to our daily work?"[8] He goes on to observe that

> race, class, and gender are the words which we see and hear everywhere at our professional meetings and in our current publications. But curiously enough [...] the English profession has failed to respond in any significant way to the issue of

---

[5] Lawrence Buell, *The Environmental Imagination: Thoreau, Nature Writing, and the Formation of American Culture* (Cambridge MA: Belknap, 1995): 2.

[6] McHarg, quoted in William Rueckert, "Literature and Ecology: An Experiment in Ecocriticism," in *The Ecocriticism Reader*, ed. Glotfelty & Fromm, 114.

[7] Glotfelty, "Introduction: Literary Studies in an Age of Environmental Crisis," xxi.

[8] Glen A. Love, "Revaluing Nature: Toward an Ecological Criticism," in *The Ecocriticism Reader*, ed. Glotfelty & Fromm, 227.

> the environment, the acknowledgment of our place within the natural world and our need to live heedfully within it, at peril of our very survival.[9]

Love's implication that a critical engagement with questions of race, class and gender are incompatible with concern for the environment is a curious assumption, and one to which I return. While he leaves unexplained the precise mechanism by which the work of individual scholars, refracted through the profession of literary studies, might effect changes at the political level, the substance of Love's critique, as with the others I have mentioned, seems to be shaped by the general question posed by Buell: "must literature always lead us away from the physical world, never back to it?"[10]

The answer, for ecocritics, must clearly be no. As to the path by which literature might lead back to the physical world, this is where Rueckert's second epigraph comes in.

> "Any living thing that hopes to live on earth must fit into the ecosphere or perish."
> — Barry Commoner, *The Closing Circle*

Logically prior, though, it would seem, intuitively secondary, to a political emphasis on the power of ecocriticism to *change* the world is an epistemological assumption concerning the means by which ecocriticism might *know* the world. "The world," from an ecocritical perspective, is "an immensely complex global system, in which energy, matter, and ideas interact."[11] Accordingly, an approach to literature which abstracts it from its embeddedness in a largely non-human material environment – i.e. most traditional forms of criticism – can no more hope to know or understand literature than a scientific interpretive framework which fails to take culture into account can hope to understand the material environment. For this reason, ecocriticism is committed to an interdisciplinary perspective which enhances the focus of literary criticism by recalibrating it with the insights of science, in particular the science of ecology.

As to exactly how ecology might work in this context, there is little agreement among ecocritics. Much contemporary ecocriticism operates according to a variation of the approach set out by Rueckert, which uses concepts from ecology – systems, energy transfer, and interdependence – in order to

---

9 Love, "Revaluing Nature," 226.

10 Buell, *The Environmental Imagination*, 11.

11 Glotfelty, "Introduction: Literary Studies in an Age of Environmental Crisis," xix.

explain the way in which literature functions in the world. Some of these ideas – such as Rueckert's suggestion that seeing a poem as "stored energy (as active, alive, and generative, rather than as inert, a kind of corpse upon which one performs an autopsy [...]) frees one from a variety of critical tyrannies, most notably, perhaps, that of pure hermeneutics, the transformation of this stored creative energy directly into a set of coherent meanings"[12] – sound a bit like jumped-up versions of New Criticism. The difference between New Criticism and ecocriticism, however, is analogous to the difference between Linnaean natural history and post-Darwinian evolution – the former emphasizes organic stability and the integrity of the individual species while the latter emphasizes change and interconnectedness. The key word here is perhaps "analogous." Some ecocriticism, such as the work of Joseph Meeker and Katherine Hayles, seeks to explain interactions between and within cultures and environments according to principles of biology and chaos theory.[13] Yet there is a temptation for many ecocritics, who tend not to be trained in science, simply to import such condensed ecological formulae as Barry Commoner's First Law of Ecology that "everything is connected to everything else"[14] into a literary-critical context in order to talk about everything from textual representations of environments ("are the values expressed in this play consistent with ecological wisdom?"[15] to the ways in which formal literary structures reflect and / or interact with organic physical ones.[16] More recent ecocriticism, chastened by the deconstructive energies of literary theory, tends to be less effusive about the positivistic power of poetry and more liable to focus on the inherent instability and interdependence of literature and the physical world. SueEllen Campbell, for example, highlights the parallels between ecocritical and poststructuralist challenges to

---

12 Rueckert, "Literature and Ecology: An Experiment in Ecocriticism," 110.

13 See also Chiu Chin–jung Lee, "Interdisciplinary Approaches to Literature: Toward an Ecological Hermeneutic," *Tamking Review* 23.1–4 (1992–93): 537–61, who offers a useful critical survey of various attempts to bring scientific theories productively to bear on literary texts.

14 Barry Commoner, *The Closing Circle: Nature, Man and Technology* (New York: Alfred A. Knopf, 1972): 33

15 Glotfelty, "Introduction: Literary Studies in an Age of Environmental Crisis," xix.

16 Richard Kerridge echoes Glotfelty in his assertion that ecocriticism seeks primarily "to evaluate texts and ideas in terms of their coherence and usefulness as responses to environmental crisis"; Kerridge, Introduction to *Writing the Environment: Ecocriticism and Literature*, ed. Kerridge & Neil Sammells (London: Zed, 1998): 5.

the humanist tradition, noting "theory and ecology agree that there's no such thing as a self-enclosed, private piece of property, neither a deer nor a person nor a text nor a piece of land."[17]

Of course, Campbell's statement, in its seamless integration of the concepts of intertextuality and biological interdependence, glosses over a number of contradictions, most obviously the incommensurability of deer and texts. The recognition that the relationship between matter and ideas is not explicable entirely by theories of ecology is what informs the third epigraph to Rueckert's article, which represents a model no less prescriptive than the previous two of the function of poetry.

> the function of poetry [...] is to nourish the spirit of man by giving him the cosmos to suckle. We have only to lower our standard of dominating nature and to raise our standard of participating in it in order to make the reconciliation take place. When man becomes proud to be not just the site where ideas and feelings are produced, but also the crossroad where they divide and mingle, he will be ready to be saved. Hope therefore lies in a poetry through which the world so invades the spirit of man that he becomes almost speechless, and later reinvents language.
> — Francis Ponge, *The Voice of Things*

The continuation of biological metaphor with which this epigraph begins (the image of man suckling the cosmos) and the political imperative with which it continues (we have to stop trying to dominate nature) should not mask the fundamentally *aesthetic* function of this prescription. Inspired by the relationship (ideal or actual) between the critic and the world and between the text and the critic, ecocriticism is also shaped by ideas about the relationship between the world and the text – ideas that owe as much to theories of art as they do to theories of science. That this aspect of ecocriticism is frequently played down in writing about the discipline may be attributable simply to its self-evidence: as an approach derived primarily from literary criticism, ecocriticism is obviously concerned with ideas about literary value. A less obvious but, I think, persuasive explanation may be derived from importing into the ecocritical context the terms of David Carter's critique of postcolonialism as a theory that would be affronted by the suggestion that it

[17] SueEllen Campbell, "The Land and Language of Desire: Where Deep Ecology and Post-Structuralism Meet," in *The Ecocriticism Reader*, ed. Glotfelty & Fromm, 133.

was motivated by "mere" aesthetics.[18] Carter cites as evidence of the operation of aesthetic value in postcolonial criticism the marked preference for texts which afforded the critic the satisfaction of performing a skilful deconstructive manoeuvre – self-righteously in the case of nineteenth-century colonial texts, vicariously in the case of contemporary texts which often perform their own deconstructive exercise. Strikingly absent from the postcolonial critical repertoire, Carter suggests, is a formula for reading realist texts, whose flat surfaces deny the readerly pleasures of ironic superiority or complicity. Ecocriticism, by contrast, prefers the realist text; it is clearly after a different kind of readerly empowerment, one connected, perhaps, with the values invoked by Ponge of "reconciliation," "hope" and "salvation." In fact, the transcendent literary experience to which Ponge refers is one explicitly associated not so much with the untranscendent genre of realism as it is with poetry – a poetry in which language is almost incidental to the power of nature which invades the reader's being. What the two forms have in common is their capacity to produce the illusory impression of an unmediated reflection of the world. They thus offer the aesthetic experience of congruence – not in an intra- or intertextual sense – but in the sense of a connection verging on identification between the human imagination and the physical environment. Stemming partly from ecocriticism's privileging of texts that seem to offer a straightforward conduit to the world is its unsurprising preference for texts that give thematic prominence to aspects of nature. Thus, in what he admits is a reductive checklist outlining what constitutes environmental literature, Lawrence Buell give high marks to texts in which "the nonhuman environment is present not merely as a framing device," "human accountability to the environment is part of the text's ethical orientation," and "some sense of the environment as a process rather than a constant is at least implicit"; by this criterion, he suggests, "James Fenimore Cooper's *Pioneers* is a more faithful environmental text than the four ensuing Leatherstocking Tales [...] while his daughter Susan's *Rural Hours* [...] is a more faithful text than any of her father's romances."[19] Such distinctions highlight what is perhaps the logical extension of ecocriticism's aesthetic preferences, which is its commitment to canon reform. Manifest at the simplest level as an attempt to rehabilitate nature writing in the eyes of

---

[18] David Carter, "Tasteless Subjects: Postcolonial Literary Criticism, Realism and the Subject of Taste," *Southern Review* 25 (1992): 292–303.

[19] Buell, *The Environmental Imagination*, 7–8.

the critical establishment, ecocriticism seeks also to inspire a more general debate about literary values – a debate which Glen Love hopes might result in the "reordering of the literary genres, with realist and other discourse which values unity rising over post-structuralist nihilism."[20]

Notwithstanding the slightly embarrassing effusiveness of Rueckert's last epigraph, most contemporary ecocritics do not subscribe to the naïve view of realism or romantic poetry as an actual window on the world; rather, they suggest that a culture which privileges the aesthetic experience afforded by that kind of literature is more likely to relate to the world in ecologically healthy ways than a culture that fetishizes "post-structuralist nihilism."[21] In the absence of empirical studies to show whether or not this is true, we can only draw inferences about what ecocriticism has managed to achieve in the way of literary critical and environmental reform. Notwithstanding the problems at which I have been broadly hinting and which I want to spend the remainder of this piece discussing, I think it important to acknowledge that ecocriticism *has* altered the terrain of literary criticism in some profound and useful ways. Apart from bringing to critical attention a number of texts which until now had been disdained because (to paraphrase the much-parodied explanation with which one New York publisher rejected the manuscript of Norman Maclean's *A River Runs Through It*) "they have trees in them," ecocriticism has also produced suggestive new readings of a number of canonical texts, highlighting the eclipsed role of the non-human world in works ranging from Charlotte Perkins Gilman's "The Yellow Wallpaper" to the writings of Oscar Wilde.[22] More importantly, ecocriticism has brought literary criticism productively to bear on science and vice versa in order to highlight both the cultural constructedness of scientific ways of seeing nature and the inescapable embeddedness of culture in biological processes.

---

[20] Love, "Revaluing Nature," 236.

[21] In his cogently argued essay "The (Im)possibility of Ecocriticism," Dominic Head notes the limitations of such prescriptive approaches to genre, suggesting that ecocriticism must, if it is to intervene productively in contemporary debates about literary studies, respond to the implications of post-modernity, even if "such a response would seem to necessitate a compromise on ecocentric values" (hence ecocriticism's [im]possibility); Head, "The (Im)possibility of Ecocriticism," in *Writing the Environment*, ed. Kerridge & Sammells, 38.

[22] See Lee Schweninger, "Reading the Garden in Gilman's 'The Yellow Wallpaper'," *ISLE* 2.2 (1996): 25–44, and Neil Sammells, "Wilde Nature," in *Writing the Environment*, ed. Kerridge & Sammells, 124–33.

It is perhaps churlish to observe that while ecocriticism has been gaining greater legitimacy, spawning readers, conferences, associations and academic positions – even quasi-recognition by the MLA – environmental problems have been getting worse. To hold ecocriticism responsible for environmental degradation is of course absurd, even if one took seriously some of the claims for its political efficacy. Nevertheless, I would maintain in the background, without dismissing outright, a connection between ecocritical success and environmentalist failure while focusing in more detail on what would appear to be a side issue – an apparent failure in ecocriticism's model of inclusivity and interconnectedness. Here the spectres of race, class and gender, which Glen Love sought to dispel, return to haunt the ecocritical scene.

Notwithstanding their alleged ubiquity in other critical fields, race and class, in particular, are conspicuous in ecocriticism mostly by their absence.[23] In her largely celebratory account of ecocriticism in the Introduction to *The Ecocriticism Reader*, Glotfelty is still moved to acknowledge that ecocriticism remains "predominantly a white movement," located primarily in the USA. She goes on confidently to predict, however, that as ecocriticism becomes "ever more interdisciplinary, multicultural, and international," it will also inevitably develop into "a multi-ethnic movement when stronger connections are made between the environment and issues of social justice, and when a diversity of voices are encouraged to contribute to the discussion."[24] Infused with the hopeful language of "connections" and "diversity," Glotfelty's diagnosis of ecocriticism's problems, and her prescription for their cure, seem to stem from the premise that the failure is one of exclusivity and narrowness. Her proposed solutions of interdisciplinarity, activism and internationalism all argue essentially for the extension of ecocriticism — from literary criticism to science, from the academy to the political arena, and from the USA to the rest of the globe. Literature – and by extension, ecocriticism, must lead back to the world, in a way that highlights and embodies the ecological law that "everything is connected to everything else." I want

[23] Recent efforts to remedy this absence include essays by Paula Willoquet–Maricondi, "African Animism, *Négritude*, and the Interdependence of Place and Being in Aimé Césaire's *A Tempest*," *ISLE* 3.2 (96): 42–62, Gretchen Lengler, "Body Politics in American Nature Writing. 'Who May Contest for what the Body of Nature will be?'," in *Writing the Environment*, ed. Kerridge & Sammells, 71–88, and, though it approaches issues of race and class indirectly, Lawrence Buell, "Toxic Discourse," *Critical Inquiry* 24 (1998): 639–65.

[24] Glotfelty, "Introduction: Literary Studies in an Age of Environmental Crisis," xxv.

to suggest that the failure of ecocriticism to address issues of race and class is not, as Glotfelty would have it, a glitch in ecocriticism's otherwise successful adherence to that principle; rather it is symptomatic of a fundamental flaw in the principle of ecocriticism itself. In order to examine the flaw, it is necessary to look more closely at the nexus of interdisciplinarity, internationalism and activism, all of which are identifiable in the principles of ecocriticism outlined by Rueckert.

The call for interdisciplinarity proceeds partly from the recognition that, in Glotfelty's words, "the environmental crisis has been exacerbated by our fragmented, compartmentalized, and overly specialized way of knowing the world."[25] Thus Glotfelty and other ecocritics call for a more holistic way of knowing, in the hope that, as Rueckert puts it "that old pair of antagonists, science and poetry, can be persuaded to lie down together and be generative after all."[26] What is notable, and what is also problematical, about framing the intersection of science and the humanities in this way is that, sanctioned by the often unspoken aesthetic emphasis on mimeticism, it subordinates the principles of interdisciplinarity to those of ecology, and not the other way around. This is a problem not because the intersection of disciplines cannot be defended on ecological grounds but, rather, because the recognition of their disciplinarity cannot be understood according to the logic of ecology. Ecocriticism's construction of interdisciplinarity, by focusing primarily on the holistic or generative potential of science and literary theory, risks falling into the trap of seeing them only as lenses through which to look at things, and not as institutionalized ways of seeing with *histories*. Ecology's history is particularly salient here.

Ecology and Globalization

The British biologist E.B. Worthington has described the twentieth century as "the ecological century." It is an accurate statement, insofar as ecology, though established in the nineteenth century, has achieved significant stature in both science and political discourse, in tension and conjunction with major movements of the twentieth century – environmentalism, most obviously, but also decolonization and globalization. The task of analysing ecology as a science and / or a philosophy, let alone tracing its history, is

25 Glotfelty, "Introduction: Literary Studies in an Age of Environmental Crisis," xxii.

26 Rueckert, "Literature and Ecology: An Experiment in Ecocriticism," 107.

complicated by its contradictory antecedents and aims. Itself an interdisciplinary science, ecology is further fragmented by conflicting ideas about whether it is or should be pure or applied, descriptive or prescriptive, conflicts which are themselves partly rooted in a philosophical opposition between what Donald Worster describes as imperial (reductionist, mechanical) and arcadian (holistic, organismic) ideologies.[27] It is in part the very contradictoriness of ecology which has rendered it susceptible not so much to misreading as to wholesale appropriation into different contexts; in fact it is debatable to what extent ecology, as a science which seeks to understand "the structure and function of nature,"[28] can be abstracted from the multiple uses to which it has been put.

The association of ecology with empire is one that has been comprehensively outlined by environmental historians.[29] On the simplest level, the use of ecology in the colonies sprang from a pragmatic need to understand problems such as soil erosion and pest control in what were, to Europeans, unfamiliar environments. On a more abstract level, the perspectival shifts that accompanied colonialism, which involved reconfiguration of space and time, with particular emphasis on man's and other species' historical adaptation to different environments, required a new construction of the physical world. Shaped by the concepts of Darwinian evolution, ecology's construction of nature as "a set of intricately connected systems that could only be understood through quantitative studies of complex interactions among species and with the land,"[30] answered the need for a perspective on change and interdependence which Linnaean taxonomy could not provide. As an interdisciplinary science which was regarded, for a long time, with hostility by many in the scientific establishment, ecology may have offered a secon-

---

[27] Donald Worster, *Nature's Economy: The Roots of Ecology* (Cambridge: Cambridge UP, 1994). This opposition (to which, it should be noted, some ecologists object – see, for example, Gregg Mitman, review of *Nature's Economy: A History of Ecological Ideas*, by Donald Worster, *Isis* 86.3 [1995]: 466–67) forms an interesting analogue to opposing aesthetic preferences among ecocritics for realism or poetry.

[28] Eugene P. Odum, *Fundamentals of Ecology* (Philadelphia PA: W.B. Saunders, 1971): 3.

[29] See Richard H. Grove, *Green Imperialism: Colonial Expansion, Tropical Island Edens and the Origins of Environmentalism 1600–1860* (Cambridge: Cambridge UP, 1995), , ed. *Ecology and Empire: Environmental History of Settler Societies*, ed. Tom Griffiths & Libby Robin (Edinburgh: Keele UP, 1997), and Thomas R. Dunlap, "Ecology and Environmentalism in the Anglo Settler Colonies," in *Ecology and Empire*, ed. Griffiths & Robin, 76–86.

[30] Dunlap, "Ecology and Environmentalism in the Anglo Settler Colonies," 77.

dary psychological appeal to new nations keen to establish their independence – financially and intellectually – from First-World metropolitan science. If the former was impossible, the latter was at least suggested by ecology's function as a "vernacular and democratic science,"[31] one which spread rapidly from the agricultural colleges of the American Midwest, where it had taken comfortable root at the end of the nineteenth century, to other burgeoning nations. Tied first to the imperative of increasing agricultural yield, the focus of ecology had, by the 1940s, shifted to the needs of conservation, in response to recognition of problems associated with soil erosion, water shortages and deforestation.[32]

From the 1950s onwards, according to most accounts, ecology underwent a process of radicalization and popularization, in which it served less as a conservative instrument of national government policy, than as a subversive instrument by which to highlight problems with national development goals. In the 1960s, with the publication of books such as Rachel Carson's *Silent Spring*, ecology was taken up (with more or less active co-operation by individual ecologists) as a blueprint for grassroots environmentalist movements. Thus Howarth notes that, after 1960, "ecology advanced from description to advocacy."[33] While there is some truth to this narrative, Howarth's comment masks the extent to which ecology had, as a kind of "applied interdisciplinary biology" directed towards the end of national development,[34] always performed a kind of advocacy role. What changed was merely the ends towards which such advocacy was directed – and even here the change has in some ways been not so great as it might initially appear: the contemporary image of the ecologist as a rugged iconoclast working against a consumerist establishment – an image of which ecocritics are particularly fond – tells only part of the story.

By the 1950s, the stereotypical image of the ecologist as a fellow with "some pieces of string and a pH meter in his back pocket" had given way to the promotion of large-scale, multidisciplinary projects designed around the

---

[31] William Howarth, "Some Principles of Ecocriticism," 73.

[32] See: Grove, *Green Imperialism*, 264–308; Libby Robin, "Ecology: A Science of Empire?," J.M. Powell, "Enterprise and Dependency: Water Management in Australia," and Michael Williams, "Ecology, Imperialism and Deforestation," in *Ecology and Empire*, ed. Griffiths & Robin, 63–75, 102–24 and 169–84 respectively.

[33] Howarth, "Some Principles of Ecocriticism," 74.

[34] Robin, "Ecology: A Science of Empire?" 65.

newly developed interpretative framework of the ecosystem.[35] The international move towards systems ecology, characterized by the use of mathematical and cybernetic models, and informed by concepts of energy flows and exchanges, had a more than incidental connection with the scientific and economic imperatives of Cold War military technology. One over-enthusiastic American ecologist used a curiously mixed military / ecological metaphor in his suggestion that the new systems ecology was like a "beach head, or a moving dune, rather open for the colonization by those invading scientists who can endure the stress of an open, turbulent environment."[36] That "open, turbulent environment" assumed global proportions with the establishment of so-called Big Science endeavours such as the International Biology Project (IBP), which sought to synthesize data from all over the world into a universal ecological theory. Launched at the instigation of European and British biologists in 1959, the IBP eventually came to be dominated by the USA, and derived most of its funding from the US Atomic Energy Commission until it was given Congressional support in 1968. Described by McIntosh as a "minor league Manhattan Project,"[37] the IBP received about US$27 million from 1968 to 1974 for variously defined ecosystem studies (which included studies of radiation effects) in addition to specific programs on insect pests, genetic material, conservation and human adaptability. What was, according to one ecologist, "in all probability the single most important event for US ecology in the last thirty years"[38] attracted considerably less enthusiasm and funding in the UK and Australia; John Turner, one of the Australian Academy of Science's representatives at the

---

[35] Robert McIntosh, *The Background of Ecology: Concept and Theory* (Cambridge: Cambridge UP, 1985): 205.

[36] Quoted in McIntosh, *The Background of Ecology*, 214. This militaristic characterization of ecology forms an ironic counterpart to Donald Worster's suggestion that "the Age of Ecology began on the desert outside of Alamogordo, New Mexico on July 16, 1945," with the detonation of the first atomic bomb; Worster, *Nature's Economy: The Roots of Ecology* (Cambridge: Cambridge UP, 1994): 342. The traditional association of ecology with anti-nuclear politics tends to gloss over its implication in the research that supported subsequent tests. The contradiction of ecology's role as scientific ammunition against and information source for nuclear testing – a contradiction experienced personally by many ecologists who had to compromise their personal beliefs in order to secure funding for their research – is acknowledged by Worster, 359–64.

[37] McIntosh, *The Background of Ecology*, 235.

[38] Burgess, quoted in McIntosh, *The Background of Ecology*, 214.

Paris Meeting to set up the IBP in 1964, wrote in his report that "the subject of Biology is not yet ripe for an International Programme."[39] While Turner appeared to be right, in that the IBP failed to produce an overarching ecosystem theory,[40] the dazzling prospect of decoding nature's laws on a worldwide scale remained, to be translated into contemporary versions of global ecology.

For many, the possibility of imagining ecology on a global scale was inaugurated by the first shots of the Earth from space, depicting the planet as a big blue marble and inspiring variously ideas about its beauty, its singularity (we're all in this together), its vulnerability and its capacity to be managed.[41] "Today's ecology," Wolfgang Sachs suggests, is in the business of saving nothing less than the planet."[42] The complexity of that business is highlighted by the coincidence, observed by McIntosh, that at the time ecology was first hailed as "subversive" science by Sears in 1964, it was increasingly being adopted by the political mainstream.[43] Without minimizing the differences in

---

[39] Bruce Robbins, letter, *PMLA* 112.5 (1997): 1135.

[40] The ecosystem concept has been challenged in recent decades, along with the framework of "Big Biology" which supported projects such as IBP, in part because the systems under investigation were seen to be too large and too complex to yield verifiable empirical data. More significantly, ecology has been influenced by the broader trend, evident in other scientific developments such as chaos theory (not to mention the philosophical move towards post-structuralism) towards emphasizing the heterogeneity and disorder that underlie all apparently homogeneous systems. While the acknowledgment of heterogeneity may have led, in ecology as in other fields, to a greater emphasis on the local, it has not, arguably, diminished the imperative to produce a global theory in which individual variances are subsumed under a general model of complexity or "ordered disorder." See N. Katherine Hayles, *Chaos Bound: Orderly Disorder in Contemporary Literature and Science* (Ithaca NY: Cornell UP, 1990): 209–35, on the tension between globalizing and localizing impulses in sciences and the humanities.

[41] See Timothy W. Luke, "On Environmentality: Geo-Power and Eco-Knowledge in the Discourses of Contemporary Environmentalism," *Cultural Critique* 31 (1995): 57–81, for discussion of the ways in which the technological production of new forms of "eco-knowledge" facilitates the extension of "geo-power" on a global scale.

[42] In *Global Ecology: A New Arena of Political Conflict*, ed. Sachs (London: Zed, 1993): 17.

[43] McIntosh, *The Background of Ecology*, 1. The evidence that grassroots environmentalism and corporate-style ecology developed simultaneously flies in the face of the prevailing assumption that the movement spawned by renegades like Rachel Carson has recently been co-opted by government and big business. While the sense of betrayal evoked by the belief that "once, environmentalists called for new public virtues, now they call rather for

ecological arguments mounted by activists seeking a radical rethinking of the position of human beings as organisms within, rather than managers of, the natural world and those deployed by technocrats pursuing a policy of sustained development – arguments supported by Worster's "arcadian" and "imperial" impulses respectively – I think it is important to note the potential for the former to be co-opted by the latter and the globalizing thrust of both. The potential for the discourse of radical ecology to be incorporated into a defence of current economic practices is due not just to the assimilative capacities of capitalist discourse but also to the congruence of economic and ecological logic.[44] I am referring here not just to the notorious defence of capitalism along the lines of Social Darwinism, but also to the increasing tendency to understand the world in terms of a cybernetic system governed by complex exchanges of information and energy. The countering image of earth as a holistic organism, in which human beings are subordinated to the laws of nature, is no antidote to this picture – it just masks its totalizing implications under the rhetoric of connections and community.

---

better management strategies" (Sachs, *Global Ecology*, xv) might work to good rhetorical effect in advancing a necessary critique of managerialism, it is based on a myth of lost ecological innocence which, by the 1970s, already appeared tenuous. While the claim advanced in 1970 by Katherine Barkley and Steve Weissman, "The Eco-Establishment," *Ramparts* 8 (May 1970): 56, that "the big business conservationists and their professionals didn't buy off the movement; they built it" might be read as reductive to an (opposite) extreme, it does at least go some way towards acknowledging the historical conjunction and overlap between local and global, progressive and conservative, conceptions of ecology.

[44] For the greater part of the twentieth century, this congruence found expression in the one-way importation of economic models into ecological science. "Economics," Donald Worster suggests, "took nothing from ecological biology that might have made it more aware of the environmental limits to man's industrial growth"; Worster, *Nature's Economy*, 294. The preceding two decades have seen a greater cross-fertilization of concepts, most concretely in the ideal of sustainable development, but also on a figurative level, with the proliferation of ecological metaphors in business discourse. See, for example, Paul Hawken, *The Ecology of Commerce* (New York: HarperCollins, 1994), who argues that profitability and ecological consciousness need not be mutually exclusive. In a different vein, James F. Moore's prize-winning essay "Predators and Prey: A New System of Competition," *Harvard Business Review* 71 (1993): 75–86, takes Social Darwinism to new and alarming heights, arguing that, "from an ecological perspective, it matters not which particular ecosystems stay alive; rather, it's only essential that competition among them is fierce and fair – and that the fittest survive" (86).

Guy Beney critiques the circulation of both images in such global environmental forums as the 1992 Rio Conference as part of a project to harness poor "Southern" nations into a "Northern" framework of ecological conservation subordinated to economic development. The two goals are fused together under the rubric of a global ecosystem ecology which

> dissolves the classic distinction between nature and the artificial within the notion of the 'system'. Everything becomes a system, from a galaxy to a company, from a cell to the planet, and so on. In this new framework, 'eco' [which originally referred to the local, "habitat"] is translated into a wild system, and the Western dynamics is revealed as the techno-natural pursuit of biological evolution.[45]

The easy assimilation of ecological concepts into the discourse of the "natural economy" makes a parody of Commoner's first law of ecology, as what matters in global economics is "the connectedness (or 'connectics') – anything which will reduce distances, delays, energy costs, and increase speed, information, productivity."[46] The incapacity of structures based on this logic to take into account issues of cultural sovereignty and social justice cannot be overcome by more holistic ideas of ecology, whose concern with the health of the biosphere that sustains *all* life, human and otherwise, overrides consideration of political structures. If, in ecology, "there is to be no interposing mechanism between man and man, man and thing and man and nature," as Anna Bramwell puts it, "neither must there be any wasteful, artificial state mechanisms, no bureaucracy, no unproductive 'Thing' in [William] Cobbett's words."[47]

---

[45] Guy Beney, " 'Gaia': The Globalitarian Temptation," in *Global Ecology: A New Arena of Political Conflict*, ed. Wolfgang Sachs (London: Zed, 1993): 182.

[46] Beney, " 'Gaia': The Globalitarian Temptation," 185.

[47] Anna Bramwell, *Ecology in the 20th Century: A History* (New Haven CT: Yale UP, 1989): 17. Bramwell's observation about political ecology's impatience with mediation is enlisted in support of the more tendentious claim for its natural affinity with fascism, a claim she substantiates by citing the now-infamous example of Nazi environmentalism. The extremity of this argument (an argument also advanced in slightly different form by Ferry), which would seem to demand nothing short of wholesale rejection of ecology, has unfortunately masked the legitimacy of some of its less inflammatory, more solidly grounded, criticisms.

Ecocriticism and Postcolonialism

Glotfelty's prescription for ecocriticism, to return, finally, to the principal focus of this analysis, is to transcend the 'things' that stand in the way of greater inclusivity – disciplinary boundaries, the walls of the academy, and the borders of the US nation. It is a course which, as the preceding discussion has suggested, can only entrench ecocriticism more firmly within the totalizing terms of eco-logic. I have suggested that one of the problems with this logic when it comes to issues of social justice is that it inevitably speaks in universals; it cannot recognize the operation of mediation, both in its own operation as a discourse and in the realm of culture and politics through which categories such as race are produced. In order to address this issue, it seems to me that ecocriticism needs to return to the realm from which Glotfelty and other critics want to escape: namely, that of "mere" literary criticism, to import Carter's designation into a different context. Instead of trying to "get back to the world," I suggest that ecocriticism needs to get back to theory, if it is to negotiate the difficult cultural place in which it now finds itself. It is here, perhaps, that postcolonialism can be of help.

One of the crucial insights of postcolonial theory has been to recognize the extent to which the process of colonialism was fuelled by a desire for an unmediated possession of the world – with devastating cultural and environmental consequences. Economic globalization has moved much further towards the achievement of this goal. Drawing on the insights of poststructuralism, postcolonial theory has undermined the cultural foundations of colonialism by highlighting the contradictions that inhere not just between, but also within, all putatively representational discourses, thereby pointing up the dangers of heeding claims by *any* cultural structures (including postcolonialism and ecology) to reflect the world transparently. The central impulse here, which is an impulse behind not just postcolonialism but much of humanities work, is that of critique, the goal not of arriving at knowledge, but of what Bill Readings describes as the "drawing out of the otherness of thought that undoes the pretension to self-presence."[48] Nothing, it would seem, could be farther from Buell's straightforward suggestion to get back to the world.

But if the roads between text and world, scholarship and environmental activism, seem to be blocked, there *is* a political dimension to the kind of critique I am discussing, of which ecocritics seem not to have taken account.

---

[48] Bill Readings, *The University in Ruins* (Cambridge MA: Harvard UP, 1996): 163.

I suggested that the burgeoning of ecocriticism and the decline of the environment might not be entirely unrelated. To ground the connection in a more specific example: at the same time as ecocritics call for a move away from the literary towards the scientific, and from academic theorizing to practical action, humanities programs are being decimated by conservative governments in Australia, Canada and the USA on the grounds of their alleged lack of relevance to the "real world." More generally, those governments are eliminating environmental regulations and relaxing the enforcement of existing ones, with the aim of streamlining and eliminating bureaucratic obstacles to growth. To generalize (to what some may perhaps judge an unjustifiable degree) what all these moves have in common is the dismantling of the public sector – that body which is characterized as bloated bureaucracy by conservatives, as well as by some radical ecologists,[49] but which is also the servant of a political structure which operates on the principle of representation and difference. This is difference not in the sense of the ecological metaphor of homeostatic diversity in which everyone has a voice but, rather, in a less satisfyingly direct sense in which meaning is produced through contradiction and conflict. The sphere of politics – which is also the sphere of environmental action – works through a commitment to maintaining the possibility of conflict and difference.[50] The opposite is domination, bad news not just for variously raced, classed and gendered constituents, but also for the environment.

---

[49] Edward Abbey is perhaps the best-known advocate of the eco-libertarian argument that wilderness must be preserved in part "for political reasons. We may need it someday not only as a refuge from excessive industrialism, but also as a refuge from authoritarian government, from political oppression"; Abbey, *Desert Solitaire* (New York: McGraw–Hill, 1968): 130.

[50] I am thinking of politics here in the sense employed by Slavoj Žižek, as a process of representation which "always involves a kind of short circuit between the universal and the particular; it involves the paradox of a singular that appears as a stand-in for the universal, destabilizing the "natural" functional order of relations in the social body"; Žižek, "A Leftist Plea for 'Eurocentrism'," *Critical Inquiry* 24 (1998): 988–89. The operation of politics involves a constant struggle, on the part of different particularities, to claim the territory of the universal – a struggle whose continuation depends on the possibility of repeatedly exposing the emptiness of all such claims. The viability of political struggle is threatened by the hegemony of a symbolic order – global capitalism and, potentially, global ecology – which promises to transcend the tension of representation by closing the space between the abstract and the concrete, the universal and the particular.

In suggesting that literary theory may be one way of addressing this issue, I am wary of falling into the ecocritical error of working by analogy from language to politics or, in David Carter's phrase, mistaking aesthetics for ethics. If ecocriticism is to be useful as a mode of critique, it will need to move, as postcolonial criticism has done, away from simply analysing texts to looking at the institutional structures that frame such practices with the aim, not of transcending them with spurious claims of taking it back to the streets (or the woods), but of, as a first goal, understanding how they work within, on, and through the categories of culture and environment. The critical movement which sometimes gets called green cultural studies (a movement which is related to, but in some ways crucially different from, ecocriticism),[51] is taking steps in this direction. In suggesting that ecocriticism needs to engage in a disciplinary critique, I have focused primarily on ecology; this is not to say that ecology is bad science, nor that I think it is wrong for literature and science to "lie down together and be generative." Far from pitting humanities against scientific study here, I think it is important to recognize that some areas of science have suffered a retrenchment of funding equal to or greater than that suffered by the humanities, and that the move to defend one against the other in some ways plays into the hands of those who would

---

[51] While the terms are sometimes used virtually interchangeably (see, for example, Glotfelty, xx), I would argue that green cultural studies and ecocriticism follow somewhat different trajectories, shaped to some degree by the different assumptions that shape the fields of literary and cultural studies. Head points to a distinction between the prescriptive model of ecocriticism proposed by Lawrence Buell, which privileges certain kinds of writing on the basis of a probable convergence of environmentalist concerns on the part of the author and critic, and the more broad-ranging critique offered in a text such as Andrew Ross's *Strange Weather: Culture, Science and Technology in the Age of Limits*; cited in Dominic Head, "The (Im)possibility of Ecocriticism," 38, 39 fn. 31. While Head suggests these texts might seem to exemplify two different approaches to ecocriticism, I would argue that Ross' study fits more clearly under the rubric of an environmentalist-inflected or 'green' cultural studies. Jhan Hochman sees ecocriticism as a specific movement within cultural studies defined by its "literature-based" approach; Hochman, "Green Cultural Studies: An Introductory Critique of an Emerging Discipline," *Mosaic* 30.1 (1997): 81. However, his delineation of cultural studies' Marxist foundations implicitly highlights a key difference between the two movements: cultural studies is characterized by an unease about the category of the 'natural,' which it is committed to exposing as historical. This unease, which I would defend as necessary (though, as Hochman points out, it makes it difficult to translate cultural critique into political activism) is singularly absent in much of which falls under the rubric of ecocriticism.

dismantle one or both — or see them merge happily and productively into a neat interdisciplinary entity like ecocriticism. Bruce Robbins has suggested that "it is the social welfare state, not the uniqueness of the humanities, that intellectuals should defend."[52] This need not be – indeed, cannot be – separate from the defence of the environment. It seems to me that ecocriticism can play a useful role in such a defence, but only by abandoning its quest to get back to the physical world, in favour of the more uncertain path of cultural critique.

WORKS CITED

Abbey, Edward. *Desert Solitaire* (New York: McGraw–Hill, 1968).

Barkley, Katherine, & Steve Weissman. "The Eco-Establishment," *Ramparts* 8 (May 1970): 48–56.

Beney, Guy. " 'Gaia': The Globalitarian Temptation," in *Global Ecology: A New Arena of Political Conflict*, ed. Wolfgang Sachs (London: Zed, 1993): 179–93.

Bramwell, Anna. *Ecology in the 20th Century: A History* (New Haven CT: Yale UP, 1989).

Buell, Lawrence. *The Environmental Imagination: Thoreau, Nature Writing, and the Formation of Amerian Culture* (Cambridge MA: Belknap, 1995).

——. "Toxic Discourse," *Critical Inquiry* 24 (1998): 639–65.

Campbell, SueEllen. "The Land and Language of Desire: Where Deep Ecology and Post-Structuralism Meet," in *The Ecocriticism Reader*, ed. Glotfelty & Fromm, 124–36.

Carson, Rachel. *Silent Spring* (Greenwich CT: Fawcett, 1970).

Carter, David. "Tasteless Subjects: Postcolonial Literary Criticism, Realism and the Subject of Taste," *Southern Review* 25 (1992): 292–303.

Commoner, Barry. *The Closing Circle: Nature, Man and Technology* (New York: Alfred A. Knopf, 1972).

Dunlap, Thomas R. "Ecology and Environmentalism in the Anglo Settler Colonies," in *Ecology and Empire*, ed. Griffiths & Robin, 76–86.

Ferry, Luc. *The New Ecological Order*, tr. Carol Volk (Chicago: U of Chicago P, 1995).

Glotfelty, Cheryl. "Introduction: Literary Studies in an Age of Environmental Crisis," in *The Ecocriticism Reader*, ed. Glotfelty & Fromm, xv–xxxvii.

——, & Harold Fromm, ed. *The Ecocriticism Reader* (Athens: U of Georgia P, 1996).

Griffiths, Tom, & Libby Robin, ed. *Ecology and Empire: Environmental History of Settler Societies* (Edinburgh: Keele UP, 1997).

Grove, Richard H. *Green Imperialism: Colonial Expansion, Tropical Island Edens and the Origins of Environmentalism 1600–1860* (Cambridge: Cambridge UP, 1995).

[52] Bruce Robbins, letter, *PMLA* 112.5 (1997): 1135.

Hawken, Paul. *The Ecology of Commerce* (New York: HarperCollins, 1994).

Hayles, N. Katherine. *Chaos Bound: Orderly Disorder in Contemporary Literature and Science* (Ithaca NY: Cornell UP, 1990).

——. *The Cosmic Web: Scientific Field Models and Literary Strategies in the Twentieth Century* (Ithaca NY: Cornell UP, 1984).

Head, Dominic. "The (Im)possibility of Ecocriticism," in *Writing the Environment*, ed. Kerridge & Sammells, 27–39.

Hochman, Jhan. "Green Cultural Studies: An Introductory Critique of an Emerging Discipline," *Mosaic* 30.1 (1997): 81–96.

Howarth, William. "Some Principles of Ecocriticism," in *The Ecocriticism Reader*, ed. Glotfelty & Fromm, 69–91.

Kerridge, Richard. Introduction to *Writing the Environment*, ed. Kerridge & Sammells, 1–9.

——, & Neil Sammells, ed. *Writing the Environment: Ecocriticism and Literature* (London: Zed, 1998).

Lee, Chiu Chin–jung. "Interdisciplinary Approaches to Literature: Toward an Ecological Hermeneutic," *Tamking Review* 23.1–4 (1992–93): 537–61.

Lengler, Gretchen. "Body Politics in American Nature Writing: 'Who May Contest for what the Body of Nature will be?'," in *Writing the Environment*, ed. Kerridge & Sammells, 71–88.

Love, Glen A. "Revaluing Nature: Toward an Ecological Criticism," in *The Ecocriticism Reader*, ed. Glotfelty & Fromm, 225–40.

Luke, Timothy W. "On Environmentality: Geo-Power and Eco-Knowledge in the Discourses of Contemporary Environmentalism," *Cultural Critique* 31 (1995): 57–81.

McIntosh, Robert. *The Background of Ecology: Concept and Theory* (Cambridge: Cambridge UP, 1985).

Meeker, Joseph. *The Comedy of Survival: Studies in Literary Ecology* (New York: Charles Scribner's Sons, 1972).

Mitman, Gregg. Review of *Nature's Economy: A History of Ecological Ideas*, by Donald Worster, *Isis* 86.3 (1995) 466–67.

Moore, James F. "Predators and Prey: A New System of Competition," *Harvard Business Review* 71 (1993): 75–86.

Odum, Eugene P. *Fundamentals of Ecology* (Philadelphia PA: W.B. Saunders, 1971).

Powell, J.M. "Enterprise and Dependency: Water Management in Australia," in *Ecology and Empire*, ed. Griffiths & Robin, 102–24.

Readings, Bill. *The University in Ruins* (Cambridge MA: Harvard UP, 1996).

Robbins, Bruce. Letter, *PMLA* 112.5 (1997): 1135.

Robin, Libby. "Ecology: A Science of Empire?," in *Ecology and Empire*, ed. Griffiths & Robin, 63–75.

Ross, Andrew. *Strange Weather: Culture, Science and Technology in the Age of Limits* (London: Verso, 1991).

Rueckert, William. "Literature and Ecology: An Experiment in Ecocriticism," in *The Ecocriticism Reader*, ed. Glotfelty & Fromm, 105–23.

Sachs, Wolfgang, ed. *Global Ecology: A New Arena of Political Conflict* (London: Zed, 1993).

Said, Edward. *Orientalism: Western Conceptions of the Orient* (1978; Harmondsworth: Penguin, 1991).

Sammells, Neil. "Wilde Nature," in *Writing the Environment*, ed. Kerridge & Sammells, 124–33.

Schweninger, Lee. "Reading the Garden in Gilman's 'The Yellow Wallpaper'," *ISLE* 2.2 (1996): 25–44.

Williams, Michael. "Ecology, Imperialism and Deforestation," in *Ecology and Empire*, ed. Griffiths & Robin, 169–84.

Willoquet–Maricondi, Paula. "African Animism, *Négritude*, and the Interdependence of Place and Being in Aimé Césaire's *A Tempest*," *ISLE* 3.2 (96): 42–62.

Worster, Donald. *Nature's Economy: The Roots of Ecology* (Cambridge: Cambridge UP, 1994).

Worthington, E.B. *The Ecological Century: A Personal Appraisal* (Oxford: Clarendon, 1983).

Žižek, Slavoj. "A Leftist Plea for 'Eurocentrism'," *Critical Inquiry* 24 (1998): 988–1009.

Views from Van Diemen's Land

Space, Place and the Colonial Settler Subject in John Glover's Landscapes

CATHERINE HOWELL

PAUL CARTER DESCRIBES the appropriation of Australia as, among other things, a spatial event where space becomes place, and where "culture declares its presence." The doctrine of *terra nullius*, in which "Aborigines [...] were not physically visible but they were culturally so,"[1] licensed the Europeans to fill this supposed void or space with a kind of transferred history and future.[2] In the context of a discussion about the relationships between colonialism, aesthetics and cultural identities, it can be argued that landscape painting forms a crucial symbolic and representational site on which Australian colonialism cultivated its terrain. Here, an investigation of the work of the nineteenth-century landscape artist John Glover may serve as a suitable jumping-off point for discussion of the ways in which colonial art has been made to serve particular narratives about colonialism, art history and cultural development.

In "The Chains of Arcady," the first episode of his 1975 ABC television series on Australian art, *Landscape with Figures*, Robert Hughes gives the

[1] Paul Carter, *The Road to Botany Bay* (London: Faber & Faber, 1987): xxii, xx.

[2] Ken Taylor, "Association, Landscape and Australian Sense of Place" (unpublished paper presented at *Comparing Australia*, British Australian Studies Association Conference, University of Stirling, Scotland, 28 August–1 September 1996): 2.

following introduction to the artist: "When John Glover arrived in Australia in 1831, he was 63 years old and he was a disappointed man."[3] There follows a brief and somewhat dismissive discussion in which Hughes pictures Glover as an old-fashioned and passé artist; a man who, having missed the aesthetic boat of Romanticism, retreats to the colonies in defeat. There, as a gentleman farmer, he will tend the wounds of his rejection by the Royal Academy (into which he was never elected) through the recreation in both life and art of the pastoral arcadia of his beloved classicists. Such a narrative suits Hughes, whose descriptions of Glover in *The Art of Australia* and in the later television series demonstrate the author's unfortunate attachment to what Bernard Smith terms "the fatal impact fallacy," or the notion that "it was Cook's fate to bring disaster in his wake."[4] Hughes borrows from Geoffrey Blainey, Bernard Smith and Marcus Clarke in order to construct Australia as a "paradise of dissent," peopled by malcontents, convicts and opportunists whose isolation made them prisoners of distance, rather than brave adventurers in a fecund New World.

To contemporary critics, Hughes's attempt to re-vision the history of art in Australia in light of the colonial project's failures and disappointments has a certain appeal. Nevertheless, his rhetoric does small justice to Glover's art when the latter is considered in relation to what Bernard Smith calls "the complex involutions of historical narrative."[5] On one level, this has to do with the way in which Hughes attempts to look behind the picturesque Georgian buildings, tidy harbours and park-like meadows of much early colonial art to the economic and social realities of the Australian population in the late-eighteenth and early-nineteenth centuries. For it was only in the 1830s, following the introduction of assisted immigration funded by land sales, that the number of free persons arriving in the colonies exceeded the transported.

Glover was one of these settler immigrants — and incidentally was Australia's first professional artist. He arrived in Launceston on the *Thomas Lowry* in February 1831, and six months later had purchased a property at Tea-Tree Brush, eighteen miles from Hobart, from whence he eventually moved to

---

[3] Robert Hughes, "The Chains of Arcady," written & presented by Robert Hughes, *Landscape with Figures: The Art of Australia* (Sydney: ABC, 1975).

[4] Bernard Smith, "Changing Posture," review of *Representing the South Pacific: Colonial Discourse from Cook to Gauguin*, by Rod Edmond, *Australian Book Review* 199 (April 1998): 8.

[5] Smith, "Changing Posture," 8.

Deddington in northern Tasmania.[6] Although sales of Glover's work had declined following the end of the Napoleonic Wars;[7] he had amassed considerable wealth and reputation in England as a watercolour painter and art teacher, and had been President of the Old Water-Colour Society in 1807–08[8] – hardly a position from which to retire in exile and resentment. Glover's three sons had already emigrated to the colonies, as had his friends the Allports,[9] and Glover himself looked forward to establishing himself as a substantial landowner while continuing to paint.

In a letter from Glover to his patron, Sir Thomas Phillipps, written in 1830 following the artist's decision to emigrate, he describes his enthusiasm for his prospective homeland:

> the expectation of finding a new beautiful World — new landscapes, new trees and new flowers new Animals Birds etc. etc. is delightful to me. [10]

The tone of Glover's letter may be tempered slightly by the fact that it was written partly with the aim of extracting any monies owed by Sir Thomas to the artist.[11] Nonetheless, it complements the positive reactions to the Australian landscape recorded by contemporary authors such as diarists Mary Ann Friend or Louisa Clifton,[12] keeping in mind that such a perspective must have been at least partly dependent on the securities of wealth and social position. In *European Vision and the South Pacific* (1960), Bernard Smith reports that Glover was said to have brought out £60,000, a vast sum

---

[6] Bernard Smith, *European Vision and the South Pacific 1768–1850* (Oxford: Clarendon, 1960): 195; Ron Radford & Jane Hylton, *Australian Colonial Art 1800–1900* (Adelaide: Art Gallery Board of South Australia, 1995): 56.

[7] Smith, *European Vision and the South Pacific 1768–1850*, 194; Tim Bonyhady, *Images in Opposition: Australian Landscape Painting, 1801–1890* (Melbourne: Oxford UP, 1985): 6.

[8] Bonyhady, *Images in Opposition*, 5–6.

[9] *Images in Opposition*, 6; Daniel Thomas, *Outlines of Australian Art: The Joseph Brown Collection* (Melbourne: Macmillan, 1973): 15.

[10] John Glover, letter to Sir Thomas Phillipps, 15 January 1830. Glover MS, Bodleian Library, Oxford. ANG: RAAM: Archives. Research Library, National Gallery of Australia, Canberra.

[11] Glover, letter to Sir Thomas Phillipps, 15 January 1830.

[12] Barbara Chapman, *The Colonial Eye: A Topographical and Artistic Record of the Life and Landscapes of Western Australia 1798–1914* (Perth: Art Gallery of Western Australia, 1979): 6.

which would have made him one of the richest men in the colonies.[13] Glover's exact financial situation at the time of his emigration remains unclear but, as Smith notes, he "certainly possessed considerable financial means when he arrived."[14] Unsurprisingly, the well-known artist seems to have been quickly accepted into Hobart society, appearing with his daughter at one of the Governor's levées in 1832.[15]

In a tone similar to Glover's, an advertisement for Joseph Lycett's *Views in Australia*, first published in June 1824, extols the glories of the new landscape to a British audience:

> It is, indeed, impossible to contemplate the scenes of natural grandeur, beauty, richness and variety, [with] which the colonies of NEW SOUTH WALES and VAN DIEMEN'S LAND abound, without mingled impressions of delight and wonder at such magnificent specimens of the stupendous power of Nature, as they burst upon our view in all the freshness of a new Creation![16]

However, this enthusiastic celebration of Australia "in her pristine state" quickly becomes a record of civic progress, which we might today interpret as environmental disfiguration:

> We behold the gloomy grandeur of solitary woods and forests exchanged for the noise and bustle of thronged marts of commerce; while the dens of savage animals, and the hiding places of yet more savage men, have been transformed into peaceful villages or cheerful towns.[17]

Views in Australia styled itself as "the first attempt to give the British public any adequate idea of the grandeur and beauty of [colonial] natural scenery, or any correct representation of their chief settlements."[18] Its author, Joseph Lycett, a Staffordshire portraitist and miniaturist, was a convicted forger who arrived in Sydney in 1814, secured a full pardon, and by 1819 was travelling widely in New South Wales and Van Diemen's Land making the sketches of picturesque scenery which formed the content of his book. *Views in Australia*

---

[13] Smith, *European Vision and the South Pacific*, 194.

[14] *European Vision and the South Pacific*, 194.

[15] *European Vision and the South Pacific*, 195.

[16] Joseph Lycett, *Views in Australia, or New South Wales and Van Diemen's Land.* London, 1824, repr. in *Documents on Art and Taste in Australia 1770–1914*, ed. Bernard Smith (1975; Melbourne: Oxford UP, 1990): 27.

[17] Lycett, *Views in Australia*, 27.

[18] *Views in Australia*, 27.

was published after Lycett's return to Britain, and invested the Australian landscape with a transferred English lineage of picturesque scenery, using this "mirror image" for a British audience of armchair travellers, and "as propaganda for dissemination in England to attract emigrants."[19] Glover's art could be seen as performing a similar role, and I will return to the question of the market for his paintings later. Yet, oddly enough, one of the first landscapes he completed here was as remote from a description of the particular qualities of Australian scenery as could be imagined.

Glover's *Composition, Italy* (c1831; Figure 7) is a small, Italianate landscape that could have been painted by Claude Lorrain, whose ideal landscapes Glover greatly admired. Termed one of the "English Claudes" by a contemporary, Glover had purchased two paintings by the artist at great expense and exhibited them alongside his own work in his first exhibition in London.[20] Unlike his later *View of Greenwich* (c1838), this is not a view of a specific British or European locale which might have appealed to a nostalgic colonist; it is rather an "exercise in high style,"[21] an imaginary composition including such actual Roman ruins as the Colosseum, and displaying such Claudean elements as a dark *repoussoir* of trees framing the distant view, with the landscape filtered through a warm clear light and peopled only by Italian peasants and their flocks. In contrast with Glover's English oil paintings in opaque paint, this *Composition* is executed in transparent glazes on a white ground, a technique which allowed him to capture also the intensity and clarity of the Australian light, most notably in *View of Mills Plains, Van Diemen's Land* (1833; Figure 8), a major work executed one year later.

Glover's arrival in 1831 placed him in between the "shock of the new" recorded by the surveyors, naturalists and topographical artists who first depicted the Australian landscape, and the representational conventions of the more sophisticated local market for art which emerged in the mid-nineteenth century. In *View of Mills Plains* we can see how Glover has moved away stylistically from Claude, making the foliage of his trees less bulky, removing the dark frame, and flattening the perspective. On a more Romantic note, the long, treed hill in the distance might suggest the widening of Glover's personal horizon in this new, pastoral Arcadia.

---

[19] Taylor, "Association, Landscape and Australian Sense of Place," 6.

[20] Smith, *European Vision and the South Pacific*, 198; Radford & Hylton, *Australian Colonial Art*, 56.

[21] Radford & Hylton, *Australian Colonial Art*, 58.

However, a comparison of these two works raises more interesting questions beyond the familiar narratives of artistic innovation and legitimation via analogy with a European stylistic predecessor. *View of Mills Plains* would seem to contradict that perennial theme of Australian art criticism: the idea that it was only with the development of the Heidelberg school, and late-nineteenth-century Australian impressionism, that artists in this country were able to capture truthfully the particular qualities of Australian light and landscape.[22] Yet the rhetoric of "progress" in art history dies hard. John Frow noted in his introduction to Rex Butler's *An Uncertain Smile*, that "All talk about art in the post-Romantic European world" is inscribed "within a historical dynamic: the new displaces and degrades the old."[23] Art criticism is thus seen to subscribe to a particular Hegelian "logic of innovation,"[24] a logic to which Robert Hughes reveals his allegiance by his use of such phrases as "charmingly awkward" to describe colonial art, and his description of early Australian culture as "an oxymoron [...] at least for the first 50 years."[25]

Of course, any discussion of Hughes's art history documentaries, and their various shortcomings, must acknowledge the theoretical and practical challenge for the television "high concept" documentary, which is to make the scattered dates and figures of the historical survey cohere into a compelling narrative – to make the bones speak. As Simon Rattle noted when talking of his own efforts to make a television series (*Leaving Home*, a history of twentieth-century music), art documentaries on TV have "the enormous advantage" over music histories of being able to present "a wide swathe of works rapidly. In an hour, it is perfectly possible to present fifty or sixty works of art – even a few seconds can make a point."[26] Where a series ranges rapidly over temporal and spatial territories, as do both *Landscape with Figures* and Hughes's more recent series, *American Visions*, the selection of

---

22 Cf., for example, David Hansen, *The Face of Australia: The Land and the People, the Past and the Present* (Sydney: Child & Associates, 1988): 6, and Mary Eagle & John Jones, *A Story of Australian Painting* (Sydney: Macmillan, 1994): 1.

23 John Frow, Introduction to Rex Butler's *An Uncertain Smile* (Sydney: Artspace, 1996): 5.

24 Frow, Introduction to Rex Butler's *An Uncertain Smile*, 8.

25 Hughes, "The Chains of Arcady" (1975).

26 Michael Hall, *Leaving Home: A Conducted Tour of Twentieth-Century Music with Simon Rattle* (London: Faber & Faber, 1996): viii.

works plays a peculiarly important role in authorizing the series' story as history. In Hughes's series, paintings are furnished as artefacts forming the "evidence" in support of his historical narrative; objects which by their very materiality incontrovertibly 'illustrate' Hughes' arguments.

The question of dubious representational accuracy aside, Glover's art represents far more than a visual record of one 'Pastoralist's Progress'. Indeed, in a country that was "urbanised before the landscape was colonised," Ian Burn asks: "what are artists in Australia painting when they paint 'the landscape'?"[27] This question has been answered partly by cultural mythographers such as Burn and Paul Carter, who describe the colonialist project as involving, parallel to the physical appropriation of territory and the expulsion of its original inhabitants, the process of "mapping the landscape into the imagination," and the creation of a "metaphysics for signifying difference."[28] Derrida has described this metaphysics as the "white mythology;" which, having repressed its origin in the physical and psychological violence of colonialism, becomes naturalized and, as such, is "inscribed in white ink, an invisible design."[29] The traditional critical emphasis on the truthfulness or "transparency" of the Australian Impressionists is suggestive of the presence of just such a "white mythology" in their work. Having once recovered this mythology, postcolonial critics have often attempted to look for moments in the colonial encounter when its hegemony is subverted or destabilized. For instance, Ian McLean traces how Paul Carter has celebrated the moment of "first contact," and its record in what he terms "first journey" narratives of exploration.[30] For Carter, "first journey" narratives take place in a precolonial moment, a kind of primordial time in which the symbolic order of Empire has yet to take root. Moreover, these narratives occupy a "pre-visual realm, one in which [...] lookouts (the sine qua non of picturesque touring) had still to be found."[31] Carter is less interested in the aesthetics of settlement, taking it as read that

---

[27] Ian Burn, *National Life and Landscapes: Australian Painting 1900–1940* (Sydney & London: Bay Books, 1990): ii.

[28] Burn, *National Life and Landscapes*, ii.

[29] Jacques Derrida, *Margins of Philosophy*, tr. Alan Bass (Chicago: U of Chicago P, 1982): 134, 211–13.

[30] Ian McLean, "Postscript: The Wandering Islands," in McLean, *White Aborigines* (Cambridge: Cambridge UP, 1998): 156–66; Paul Carter, *Living in a New Country* (London: Faber & Faber, 1992): 35.

[31] Carter, *Living in a New Country*, 35.

> to paint a colonial landscape of smiling fields, of hazed hill heads and glinting brooks was not to represent the lie of the land but to articulate the logic of a cultural dream and its spatial mise-en-scène. [32]

Following Carter's schema, Glover's paintings work principally to create the Australian landscape as an object of Imperial desire. Such a reading is supported by the fact that Glover's success as a colonial artist came in England, the artist finding "almost no market for his paintings in Australia"; "Glover's work passed almost unnoticed in the Tasmanian press and none of his paintings were shown at the fine art exhibitions held in Hobart in 1845 and 1846."[33] However, in 1835–36 he had exhibited sixty paintings in London at the height of the London art season, at a time when the exotic attractions of Robert Burford's panorama of the "Temple of Karnak in Egypt" and a diorama of the "Church of Santa Croce" were also competing for viewers. Two editions of the exhibition catalogue were published, and Glover sent another consignment of 35 paintings to London in January 1836, yet these pictures were not exhibited (possibly due to customs difficulties) and were distributed among members of the artist's family.[34] This proved to be the artist's last attempt to exhibit overseas.

It would seem that Glover's paintings did not fulfil the role of representing the new colony to itself, or at any rate to the wealthier merchants and landowners who desired to possess aesthetic confirmation of their personal taste and material success in the colonies. It was not until the artist's death in 1849 that his paintings were disseminated locally, when his children had them raffled off.[35] Glover was no portraitist, and neither did he illustrate grandiose historical / mythological themes or still lifes in the manner of two other prominent colonial artists, Augustus Earle and W.B. Gould. Moreover, unlike Conrad Martens or John Skinner Prout, Glover painted few views of Tasmanian settler homesteads other than his own, despite the fact that "views from and of country houses had formed a small but significant part of Glover's English oeuvre"; those he did paint – "anachronistic representations of an Aboriginal arcadia" – often exclude evidence of the colonist's achievements altogether. [36]

---

[32] Carter, *Living in a New Country*, 12–13.

[33] Bonyhady, *Images in Opposition*, 8–9.

[34] *Images in Opposition*, 9.

[35] *Images in Opposition*, 9.

[36] *Images in Opposition*, 47.

This "anachronism" is in fact the central feature of Glover's work. As has been suggested here, his paintings embody a tension between topographical art's concern for accuracy, the Italianate conventions of the ideal landscape, and a freer, more individualist, Romanticism derived from artists such as Salvator Rosa. It would be easy enough to furnish reasons for Glover's failure to paint other settler's properties, such as his age and his isolated location in Northern Tasmania. More important perhaps was his pleasure in painting his own property. Glover was a farmer's son and had begun his career as a provincial drawing master in the English Midlands.[37] Surely his many paintings of his own property, which he named 'Patterdale' after his previous home district in Cumberland,[38] would have reinforced his pride and sense of self-worth in establishing himself as a landowner and artist who could afford not to depend upon a local market (see Figure 5). For we can read the European occupation of Australia in landscape art not only as a spatial event with connotations of aesthetic value, but as being bound up with "the Lockeian principle of private property in land," and – for rural labourers as well as extant landowners – "its future production where land has a utilitarian value."[39] Paintings such as *My Harvest Home* (1835; Figure 3) and Glover's most famous work, A *View of the Artist's House and Garden, in Mills Plains, Van Diemen's Land* (1835; Figure 6), bear out this interpretation with their views of domesticated, cultivated landscapes sited within the context of the Tasmanian wilderness.

However, further investigation of critical discussion of the 'anachronisms' of colonial landscape painting often reveals a formalist agenda, quite different from Nicholas Thomas' or Tim Bonyhady's more materialist concerns. Once again, this has to do with the historical transformation of a painting into a "carcase [which] can become, eventually, a fossil carrying information into the remote future."[40] When Hughes criticizes colonial art as naive, or when Carter argues that it is ideologically overdetermined, both are referring to the relation embodied in the artworks between form and content, style and perception. In the case of both critics, the style or representational 'frame' of the colonial artwork is seen as both deficient and overloaded – both 'not enough' and 'too much'. In Carter's case, I would argue (along

[37] Thomas, *Outlines of Australian Art*, 15.

[38] Smith, *European Vision and the South Pacific*, 196.

[39] Taylor, "Association, Landscape and Australian Sense of Place," 6.

[40] Thomas, *Outlines of Australian Art*, 14.

with McLean) that the longing for a non-ideological space leads him to elide the complexities of colonial representations.[41] In Hughes's modernist trajectory, what is principally necessary is for colonial art to "grow up," and in so doing to forge an identity for national aesthetic production grounded in the knowledge of the sins and omissions of the colonialist project. Both approaches pay insufficient attention to the fact that "postcolonialism is distinguished, not by a clean leap into another discourse, but by its critical re-accentuation of colonial and anticolonial languages."[42]

This is especially pertinent in approaching Glover's complex depictions of Tasmanian Aborigines, a subject which has already been discussed at length by Tim Bonyhady and Radford & Hylton.[43] While I do not discuss this issue in depth here, I will attempt in what follows to indicate some crucial points arising in relation to the question of Indigenous subjects and landscape in Glover's art, with particular reference to *A Corrobery* [sic] *of Natives in Mill's Plains* (1832; Figure 5). This major work depicts a fiery sunset above a clearing in a gum forest, where a group of anonymous Indigenous figures enact a ceremonial gathering around a fire.

Glover frequently inserted Aboriginal subjects into his Arcadian landscapes, often depicting Indigenous people as carefree inhabitants of a fertile and benevolent land, untroubled by European settlement. European and Indigenous subjects rarely share the same space in Glover's work. In paintings such as *The Bath of Diana* (1837), and *The River Nile, Van Diemen's Land, from Mr Glover's Farm* (1837) the Indigenous subject is absorbed by the titles' classical references into the organic time of a Golden Age. At one level, this might be seen as entirely disingenuous on Glover's part. By the time the artist arrived in Tasmania, the Aboriginal population was decimated. The Black War had just ended, and the then Protector of Aborigines, George Augustus Robinson, had rounded up the two-hundred-odd remaining Indigenous people for confinement on Flinders Island in Bass Strait, where most of them rapidly died of despondency.[44] Glover made several drawings of Aborigines in 1831–32, "which together show all twenty-six members of the

---

[41] McLean, "Postscript: The Wandering Islands," 160.

[42] Nicholas Thomas, *Colonialism's Culture: Anthropology, Travel and Government* (Melbourne: Melbourne UP, 1994): 7.

[43] Bonyhady, *Images in Opposition*, 23–39; Radford & Hylton, *Australian Colonial Art*, 68–70.

[44] Radford & Hylton, *Australian Colonial Art*, 68.

Big River and Oyster Bay tribes immediately before they were sent to Flinders Island."[45] Glover's London exhibition catalogue also records that several of his paintings were based on first-hand experience, entry no. 44 stating that "these Natives Danced and Bathed at the request of the Artist" and no. 55 that "The Natives climbed the Tree to shew their method of catching Opossums."[46]

A Corrobery of Natives in Mills Plains was thus a reconstruction partly from memory, in part from sketches made early in 1831 in Hobart by Glover, and possibly also from sketches he made of the ten New South Wales Aborigines imported by John Batman to hunt and entrap the remaining local Aboriginal people.[47] Glover seems to have shown some interest in Aboriginal customs, accompanying Batman and the mainland Aboriginal people on an expedition during which he included them in landscape sketches. Glover was later to write that Batman was "a rogue, thief, cheat and liar, a murderer of blacks and the vilest man I have ever known."[48] Yet, despite his apparent sympathy with their plight, the artist's contact with the Indigenous people was inevitably superficial. Glover does not seem to have noticed the differences in material culture between Tasmanian and mainland Aboriginal people, as demonstrated by *A Corrobery*, which depicts figures with shields when these were not in fact carried by Tasmanian Aborigines.[49]

Yet *A Corrobery of Natives in Mills Plains* also demonstrates that Glover's subjects were not restricted to the "smiling fields" and "glinting brooks" referred to by Paul Carter. One element of Glover's relationship to neoclassical art that is not often commented upon is that the latter's conventions inevitably include an intimation of trouble in paradise. Potentialities of death, ruin or disappointment are present in the form of subtle emblematic suggestions: "the hunted stag, the crumbling castle are carefully placed, objects of contemplation there to be teased [...] into metaphysical significance."[50] One thinks of Poussin's famous painting *Et in Arcadia Ego* (c.1630), in which a shepherdess and her companions stumble across a tomb in the bucolic

45 Radford & Hylton, *Australian Colonial Art*, 68.

46 John Glover, *A Catalogue of Sixty Eight Pictures* (London, 1835, repr. 1868; Bray Reference Library, State Library of South Australia, Adelaide 1835).

47 Bonyhady, *Images in Opposition*, 32; Radford & Hylton, *Australian Colonial Art*, 68.

48 Radford & Hylton, *Australian Colonial Art*, 68.

49 Bonyhady, *Images in Opposition*, 32.

50 Andrew Graham–Dixon, *Paper Museum* (London: Fontana, 1997): 153.

setting. Paintings by Glover such as *A Corrobery of Natives in Mills Plains* and *The Last Muster of the Tasmanian Aborigines at Risdon* (1836) combine neo-classical melancholy with the more urgent mood of Romantics such as Salvator Rosa. As a settler and farmer, Glover is complicit in the displacement and extirpation of the Indigenous people he depicts. Yet one suspects that the theme of dispossession played uneasily on his mind, and the artist was to use the corroboree as a subject for several pictures, including his last painting, begun on his seventy-ninth birthday. On it he inscribed the same tribute he used in his 1835 catalogue and on two corroboree paintings he presented to the Musée du Louvre in 1842, "I have seldom seen more enjoyment and mirth on such occasions than I ever saw in a Ball Room in England."[51] With its dying, lurid light, weirdly snaking eucalypts and tiny figures absorbed into the shadows of the landscape, *A Corrobery of Natives* can be seen as Glover's monument to a race he believed doomed.

Recent investigations of the concept of landscape[52] have often privileged the themes of environmental awareness and a rejection of anthropocentric worldviews. An eco-centred view of landscape art might see such art as expressing an awareness of a new relationship between humanity and the world it inhabits. This landscape "is more than the extension of actual space or the scope of knowledge; it is also an acknowledgment of alienation and insignificance in the spectator, the self."[53] If the break with medieval mental structures that took place during the Renaissance focused on "the active vigorous individual who sees himself as the measure of all things," it also inspired respect for "the observation and investigation of the environment, which implied confrontation with the unknown and unobserved."[54] Glover's painting *Mount Wellington with Orphan Asylum – Van Diemen's Land* (1837; Figure 4) is the only known Tasmanian landscape by the artist that conveys a sense of the vastness of the wilderness. The orphanage, symbol of civic progress,[55] is dwarfed by the forest which surrounds it and by Mt Wellington looming in the background. Glover has chosen not to include evidence of the Hobart to Launceston highway in order to emphasize more fully the

---

[51] Quoted in Radford & Hylton, *Australian Colonial Art*, 70.

[52] Such as Simon Schama's *Landscape and Memory* (London: HarperCollins, 1995).

[53] Margaretha Rossholm Lagerlöf, *Ideal Landscape: Annibale Carracci, Nicholas Poussin and Claude Lorrain* (New Haven CT & London: Yale UP, 1990): 2.

[54] Lagerlöf, *Ideal Landscape:*, 2.

[55] Bonyhady, *Images in Opposition*, 73.

plight of the orphans; but, as Bonyhady notes, the inclusion of a rainbow over the scene signals divine protection and hope for the future. In fact, this complex blend of Christian convention, secular philanthropy and sublime landscape imagery points towards the important role played by later North American landscape artists such as Frederick E. Church in the establishment of national parks.[56]

In conclusion, study of Glover's work in light of the critical histories that have taken colonial art as their subject reveals disparate narrative trajectories and political goals. Within these competing trajectories, history is remade as popular narrative, as cultural myth or ground, and even as eco-parable. Glover's landscapes, however, are not easily assimilable into any of these analyses. Rather, his work displays a complex and, at times, contradictory attitude to the relationship between the colonial settler subject and his environment. If my reading of his work makes a more general point beyond this, it would be to reaffirm that postcolonial work in an Australian context involves not just a reappraisal of those images and stories created long before the advent of what we have learned to call Theory, but also of the traffic in received ideas that takes place post-theory, most especially in the case of simplistic constructions of the past as deluded and of the present as enlightened.

WORKS CITED

Bonyhady, Tim. *Images in Opposition: Australian Landscape Painting, 1801–1890* (Melbourne: Oxford UP, 1985).

Burn, Ian. *National Life and Landscapes: Australian Painting 1900–1940* (Sydney & London: Bay Books, 1990).

Carter, Paul. *Living in a New Country* (London: Faber & Faber, 1992).

——. *The Road to Botany Bay* (London: Faber & Faber, 1987).

——. "Second Sight: looking back as colonial vision," *Australian Journal of Art* 13 (1996). 12–13.

Chapman, Barbara. *The Colonial Eye: A Topographical and Artistic Record of the Life and Landscapes of Western Australia 1798–1914* (Perth, W.A.: Art Gallery of Western Australia, 1979).

Derrida, Jacques. *Margins of Philosophy*, tr. Alan Bass (Chicago: U of Chicago P, 1982).

[56] Robert Hughes, "The Wilderness and the West," written & presented by Robert Hughes, *American Visions*, prod. Planet 24 in association with BBC-TV, ABC, Sydney, 11 August 1996.

Eagle, Mary, & John Jones. *A Story of Australian Painting* (Sydney: Macmillan, 1994).

Frow, John. Introduction to Rex Butler's *An Uncertain Smile* (Sydney: Artspace, 1996).

Glover, John. Letter to Sir Thomas Phillipps, 15 January 1830. Glover MS, Bodleian Library, Oxford. ANG: RAAM: Archives. Research Library, National Gallery of Australia, Canberra.

——. *A Catalogue of Sixty Eight Pictures*. London, 1835. Repr. 1868. Bray Reference Library, State Library of South Australia, Adelaide.

Graham–Dixon, Andrew. *Paper Museum* (London: Fontana, 1997).

Hall, Michael. *Leaving Home: A Conducted Tour of Twentieth-Century Music with Simon Rattle* (London: Faber & Faber, 1996).

Hansen, David. *The Face of Australia: The Land and the People, the Past and the Present* (Sydney: Child & Associates, 1988).

Hughes, Robert. *The Art of Australia: A Critical Survey* (Melbourne: Penguin, 1966).

——. "The Chains of Arcady." Written & presented by Robert Hughes. *Landscape with Figures: The Art of Australia* (Sydney: ABC, 1975).

——. "The Wilderness and the West." Written & presented by Robert Hughes. *American Visions*. Prod. Planet 24 in association with BBC-TV. ABC, Sydney, 11 August 1996.

Lagerlöf, Margaretha Rossholm. *Ideal Landscape: Annibale Carracci, Nicholas Poussin and Claude Lorrain* (New Haven CT & London: Yale UP, 1990).

Lycett, Joseph. *Views in Australia, or New South Wales and Van Diemen's Land*. London, 1824. Repr. in *Documents on Art and Taste in Australia 1770–1914*, ed. Bernard Smith (1975; Melbourne: Oxford UP, 1990): 26–29.

McLean, Ian. "Postscript: The Wandering Islands," in McLean, *White Aborigines* (Cambridge: Cambridge UP, 1998).

Radford, Ron, & Jane Hylton. *Australian Colonial Art 1800–1900* (Adelaide: Art Gallery Board of South Australia, 1995).

Schama, Simon. *Landscape and Memory* (London: HarperCollins, 1995).

Smith, Bernard. "Changing Posture." Review of *Representing the South Pacific: Colonial Discourse from Cook to Gauguin*, by Rod Edmond, *Australian Book Review* 199 (April 1998): 7–8.

——. *European Vision and the South Pacific 1768–1850* (Oxford: Clarendon, 1960).

Taylor, Ken. "Association, Landscape and Australian Sense of Place," unpublished paper presented at *Comparing Australia*, British Australian Studies Association Conference, University of Stirling, Scotland, 28 August–1 September 1996.

Thomas, Daniel. *Outlines of Australian Art: The Joseph Brown Collection* (Melbourne: Macmillan, 1973).

Thomas, Nicholas. *Colonialism's Culture: Anthropology, Travel and Government* (Melbourne: Melbourne UP, 1994).

Figure 3
John Glover (born England 1767, arrived Australia 1831, died 1849), *My Harvest Home* (1835; oil on canvas, 76 x 114 cm). Collection: Tasmanian Museum and Art Gallery

Figure 4

John Glover (Britain/Australia, 1767–1849), *Mount Wellington with Orphan Asylum, Van Diemen's Land* (1837; oil on canvas, 76.5 × 114.2 cm). Purchased through The Art Foundation of Victoria with the assistance of the Joe White Bequest, Governor, 1981. National Gallery of Victoria, Melbourne

Figure 5
John Glover (Britain/Australia, 1767–1849), *A Corrobery of Natives in Mill's Plains* (1832, Deddington, Tasmania; oil on canvas on board; 56.5 × 71.4 cm). Morgan Thomas Bequest Fund 1951, Art Gallery of South Australia, Adelaide (0.1466)

Figure 6

John Glover (Britain/Australia, 1767–1849), *A View of the Artist's House and Garden, in Mills Plains, Van Diemen's Land* (1835, Deddington, Tasmania; oil on canvas, 76.4 × 114.4 cm). Morgan Thomas Bequest Fund 1951, Art Gallery of South Australia, Adelaide (0.1464)

Figure 7
John Glover (Britain/Australia, 1767–1849), *Composition, Italy* (c. 1831, Hobart; oil on canvas, 51.0 × 71.5 cm). Mrs Mary Overton Gift Fund 1990, Art Gallery of South Australia, Adelaide (901P1)

Figure 8
John Glover (Britain/Australia, 1767–1849), *View of Mills Plains, Van Diemen's Land* (1833, Deddington, Tasmania; oil on canvas, 76.2 × 114.6 cm). Morgan Thomas Bequest Fund 1951, Art Gallery of South Australia, Adelaide (0.1465)

Colonial *Cordon Sanitaire*

Fixing the Boundaries of the Disease Environment

Jo Robertson

PEEL ISLAND, off the southeast coast of the state of Queensland, is, in almost anyone's terms, an utterly insignificant island – too small to be marked on the maps of Queensland, much less Australia. It was described in a 1993 report for the Queensland Department of Environment and Heritage as "a low lying island of approximately 590 hectares." Peel Island sits in the environmental marine park of Moreton Bay, at the entrance to the Brisbane River, on which stands Brisbane, the capital city of Queensland.

If you travel from the Brisbane River out into Moreton Bay, dolphins can be seen any morning playing together in the water and huge turtles swim among the wrecks around the islands. The bay is a candidate for Paradise; Peel, in the very midst of this, represents a strange and rather pathological place of containment and incarceration, masquerading as "redemption" from fear.

In the sixteenth century, epidemic disease was understood to spread in the atmosphere: "it receives spoilt exhalations and effluvia causing disease" such as miasmas, and it "conducts them often, without amalgamating with them, for a long time and over great distances."[1] Perhaps something of the same

---

[1] Johann Peter Frank, *A System of Complete Medical Police: Selections from Johann Peter Frank*, ed. Erna Lesky, tr. E. Vilim (1788; Baltimore MD: Johns Hopkins UP, 1970): 441.

sense of taintedness continues to cling to Peel Island, for Peel was the site of a lazaret from 1907 to 1959 and, conceivably, this same sense of taintedness has shielded it from later attempts to "develop" it.

When the lazaret was closed, the island became vacant Crown land. In 1960, the government invited applications for a lease of the island for tourist development. Only one proposal was submitted. A Californian doctor wanted to construct a ring road around the island, an airstrip, a golf course, a hotel and a wildlife sanctuary. Then in 1964, there were government plans to set aside 145 acres for residential development and 27 acres for a tourist resort, with the rest as a national park, but it still failed to interest developers. In 1967, another attempt to develop the island as a tourist resort failed. Then in 1989, almost by default, the island was proposed as an Environmental Park, and is now managed from the mainland.[2]

The State Department of Environment and Heritage, with funding from the Commonwealth government, commissioned a study of Peel in 1993, in order to prepare a Conservation Plan for the buildings that remained as visible traces of the lazaret. The resulting report provides an overview of its history and an analysis of its cultural significance.[3]

Leprosy was first noticed in the colony of Queensland in 1855 in a Chinese labourer. Isolated cases occurred in the decade 1860–70, and in the 1880s it became increasingly noticed among the Chinese and Pacific Islanders. In 1891, a white man with leprosy was discovered and a bill was passed making it legal to detain and isolate those diagnosed with the disease. Various island sites were used for varying periods as detention centres for lepers. In 1891 an area at Dunwich on Stradbroke Island (a larger island adjacent to Peel Island) was set aside, and this site was used for those diagnosed with the disease until 1910. At the same time, Friday Island (in the Torres Straits) was set aside for those of "other races" from 1892 until 1910. Then in 1906, a "modern" lazaret was erected in the north-western corner

---

[2] Robert Riddel, *'The Leper shall Dwell Alone': Peel Island Lazaret Conservation Plan: A Report for the Department of Environment and Heritage* (Brisbane, Queensland: Queensland Department of Environment and Heritage, 1993): 24–25.

[3] Riddel, *'The Leper shall Dwell Alone'*, Introduction: i. The report is written under the name of Robert Riddel, a Brisbane architect, but it acknowledges the assistance and advice of a number of others: Thom Blake (who did much of the actual research and writing), the anthropologist Dr Anne Ross, an oral historian, Peter Ludlow, Dr Eric Reye, one of the last doctors on the island, and June Berthelsen, one of the sufferers of Hansen's disease who had been detained on the island.

of tiny Peel Island. There in July 1907, 71 patients were admitted: 16 Europeans, 3 Chinese, 1 Indian, 4 Aboriginals, and 47 Melanesians. From 1907 to 1959, when it closed, a total of 400 people were admitted, the most being 84 in 1910. Two hundred and fifty died in the lazaret.[4] Then after 1940 (and until the 1970s) Fantome Island, next to Palm Island, near Townsville, became the leper asylum for Queensland.

Peel Island is described in the Conservation report as having state and national significance because of all the lazarets in Australia it is the only fully segregated one, in which whites were separated from coloureds, males from females, the more diseased from less diseased, and individuals from each other: "Peel Island, therefore is significant as the only surviving example of a substantially intact lazaret that was designed on the isolation principle."[5] The site is also significant because it provides a clear illustration of the difference in treatments provided for what were perceived as different racial groups: "The stark difference in accommodation for whites and coloureds also illustrates how concepts of race influenced medical treatment in the twentieth century."[6] In addition, the lazaret itself reveals attitudes to leprosy. Underpinning all of this is the social significance of the site, for it indicates an administration dedicated to "institutions of reform, discipline, and social control."[7]

But the report also notes that, apart from its historic value, the lazaret has "unusual aesthetic qualities":

> These qualities derive from several different aspects of the site. First, there is the acute sense of the past evoked by the abandoned, decaying state of many of the buildings, the loose materials and the numerous items and artefacts scattered throughout the site.
>
> A second aspect is the sense of remoteness and seclusion of the site. This sense derives partly from the way in which the site is accessed from the other side of the island along tracts through native forest. The seclusion of the site is accentuated by the surrounding vegetation of tall native forest, the grove of mango trees and the mangroves along the shoreline. The aesthetic qualities of the site also include its striking picturesque attributes. The picturesque setting derives from the carefully maintained lawns in the centre of the lazaret, the

4 Riddel, '*The Leper Shall Dwell Alone*', 14.

5 '*The Leper Shall Dwell Alone*', 30.

6 '*The Leper Shall Dwell Alone*', 30.

7 '*The Leper Shall Dwell Alone*', 28.

> gardens and trees (particularly a large poinciana tree) around the superintendent's quarters, and the views to Moreton Bay.
>
> The juxtaposition of these various attributes – the sense of abandonment and decay, the seclusion and remoteness of the site, the picturesque landscape and seascape – contribute to the cultural significance of the site. [8]

In the terms of the report, the lazaret has historical and aesthetic value: decay and remoteness are hedged in by native forest, fruit trees and mangroves. Clipped lawns and gardens are surrounded by views of the Bay. In a series of concentric circles, the lawns and gardens and the decaying, remote and isolated site can be envisaged. At the very centre of these concentric circles, the built environment is slowly surrendering in palpable silence to nature, and this very decay is what the report wants to preserve – to somehow freeze the last phase before complete disintegration, before a final return to the earth, and forgetfulness. The report recommends a delicate programme of preservation, rather than restoration, of maintenance and repair rather than reconstruction. The primary quality which the report desires to preserve is the "decaying abandoned character" because the decay and encroachment of the natural vegetation "clearly indicates that the site no longer functions as a lazaret, nor has for a considerable period of time," and it recommends strict control of the site, to ensure its preservation. Peel's dual identities, its historical identity as a site of quarantine – as part of the disease environment – and its contemporary identity as conservation site, in the midst of an environmental marine park, find greater resonance when they are considered as part of a narrative of national identity

At the time of the discovery of leprosy in Queensland in the 1890s, the disease – already freighted by the late-nineteenth century with a multiplicity of connotations that conjured threats to the defining boundaries of the body – had been revivified by the new germ theory. *Bacillus leprae* had been shown, by G.H. Armauer Hansen in 1873, to be the bacillus consistently present in the nodes of leprosy patients, and the very first bacterium to be discovered in the human being.

Worldwide anxiety about the effects of the migrations of large populations was transmuted at the end of the nineteenth century into quarantine measures and legislation to control disease. Australia and, more specifically, the colony of Queensland imagined itself as extremely vulnerable to diseases that could only arrive from tides of immigration. Although ideas of quaran-

---

[8] Riddel, '*The Leper Shall Dwell Alone*', 31.

tine were hotly debated in the nineteenth century, they were enforced in the Australian colonies because they provided a sense of security as a counterpart to the imagined fragility of the health of the colony.

Quarantine was considered a necessary measure "as a preventative of invasion"[9] and while the English doctors had decided against quarantine, arguing that it only served to spread disease because it provided a false sense of security and mitigated the need to develop hygiene and sanitation measures in the cities, the *Australian Medical Journal* argued that the unique conditions in the colony made it necessary. They concluded that "it is of paramount importance to neglect no step to keep out the bacillus, and quarantine, though naturally not absolutely safe, is safer for this purpose than medical inspection."[10]

The best measure that the colony could adopt to deal with leprosy was an intensified form of quarantine.[11] Sanford Jackson, the resident doctor for the Brisbane Hospital, was asked by the Colonial Secretary for suggestions for dealing with cases. He wrote that "while it may be expedient on occasion to admit and isolate a case in a tent on the hospital grounds – popular prejudice is justly against them being retained." He suggested a temporary settlement in some quarantine ground near Brisbane, such as "a tent at Dunwich, surrounded by a pallisading fence" where "patients could be locked in, and food supplies left at the gate until transport to Torres Strait is arranged."[12] The doctor was suggesting that this disease could only be contained behind a fortress on an island, a prelude to the commencement of a form of quarantine which would become a life sentence for leprosy sufferers.

In the specific context of colonial Queensland, anxieties about degeneration in the tropics and degeneration of the racial type heightened the hysteria surrounding the question of control of individual diseased bodies. Anxiety about colonial identity was combined with economic anxiety in a debate over the reintroduction of kanaka labour. In this debate, economic survival was balanced against fears about the possible consequences of a change in ethnic composition and subsequent degeneration in the colony. One of the key issues was the racial composition of the colony: would the reintroduction of kanaka labour mean that the colony was predominantly European or

9 "Public Health in New South Wales," *Lancet* (10 October 1891): 827.

10 "Presidential Address," *Australian Medical Journal* (15 February 1893): 74.

11 "Leprosy in New South Wales," *Australasian Medical Gazette* (June 1891): 280.

12 *QSA* Col 264: No. 02195 "Response" 24 February 1891.

not? There were already perceptions of a growing Asian population representing a threat to the survival of the European races.

These tensions were aggravated by a sense of vulnerability and fear of degeneration of the European races, who were also seen to be at the mercy of changes produced by colonial acclimatization. Alarmingly, perhaps, for those in the north, the *Daily Telegraph* reproduced an essay by D. Christie Murray which discussed the effect of climate and environment on race. In the process of an optimistic forecast of the evolution of a "new racial type" ("taller, slimmer and more alert") in the nation generally, it predicted that "In the northernmost parts of Australia, it is evidently impossible that any race of men can for many generations preserve the characteristics of European peoples."[13] The prospect of degeneration for those living in the tropical climate of northern Queensland was a subject for speculation and an obvious source of anxiety. These concerns only veiled a more insidious threat – that of miscegenation; mixed-race marriages would, it was asserted, produce a degenerate type. It was also feared that the descendants of the early European immigrants were "enfeebled and infertile."[14] Whether through the effects of interracial breeding or in consequence of the climate, the very possibility of the existence of whites in the North seemed to be threatened.

In 1895, in the midst of debates about the re-introduction of indentured labour into the colony as an economic imperative, a woman named Bella Clarke was exiled to Peel Island, eleven years before it officially became a lazaret. Her potential to reduce the community to hysteria was not at first apparent; but, as her situation became public, she was progressively reconfigured until she and the island came to be considered a threat to the individual, to the family, and to the nation.

A twenty-three year old woman diagnosed with leprosy, Bella Clarke, first appears in the records in a medical report from Dr Lyons.[15] Lyons made two examinations: one at the General Hospital on 27 September 1895 and the other at the Immigration Depot on 30 September. His report was framed as an attempt to understand how she had become infected; and it contained an "account of her life" as she had related it and is a record of her sexual history and a history of her associations with indigenous people,

---

[13] D. Christie Murray, *Daily Telegraph* (10 October 1891): 1.

[14] "Leprosy in New South Wales," *Australasian Medical Gazette* (June 1891): 280.

[15] "Report from Dr Lyons on Isabella Clarke," *QSA* 30 September 1895 (Col 272: No. 11702) received by the Colonial Secretary on 1 October 1895.

Chinese and other immigrants. Initially, a genealogy of health was compiled from the medical history of her father, brothers and sister, and the phrases used by Lyons retain something of the tone of her voice. She told the doctor that her father died four months before she was born and, although she believed that he had been healthy, she really was not sure. Her brothers and her stepsister were all healthy. She left school when she was twelve in order to help her mother run a boarding house. Her indignation is registered at the suggestion that she may have associated with men of other racial groups: "She never at any time associated with Chinamen, or coloured men and women, of any kind," and she "always had an antipathy to coloured people." Her sexual history follows in a more subdued tone: she "got into trouble with a young man, and had a son to him who was consumptive and died when 12 months old." If there was a sense of tragedy associated with this, the clinical tone of the reporting erases it. Her subsequent employment as a servant, however, marks a convergence of sexual history and a history of associations with potential sources of infection. At this point, the doctor's clinical tone is occasionally disturbed by the intrusion of Bella's voice when her "association" with "blackfellows" is suggested:

> In 1893, she went up country, as a general servant, to Hornet Bank Station, 33 miles from Taroom. There were 2 blackfellows at the Station who used to have their meals in the kitchen, [she] never associated with them in any way. While here [she] became intimate with a stockrider, a whiteman and a Jew, but did not become pregnant.

Bella's narrative emerges at the intersection of discourses of medicine, sexuality and race. Her movements were mapped from Hornet Bank to Taroom and back to Brisbane, where her mother was shocked at the change in her facial features because "her nose had become flattened and her face discoloured, and [...] she had no eyebrows and eyelashes." Potentially connected to her life story is another narrative – of those whom she may have contaminated. On her return to Brisbane, she worked as a cook in a lady's house and in the five weeks leading up to her visit to the doctor she helped a neighbour with her housework. Her ill health led her to visit the General Hospital where Doctor Jackson diagnosed her as a "suspected leper" and presumably reported her. A medical report followed which mapped the metamorphosis of her body. The changes to the skin of her forehead, nose, chin and ear lobes, and the loss of eyebrows and eyelashes, the condition of her arms and her thighs, ankles and feet were sufficient for Lyons to conclude that she had "skin leprosy in a somewhat incipient stage."

Finally, she was mapped as a bacteriological specimen. On 30 September 1895, C.J. Pound, the Government Bacteriologist, sent a bacteriological report to the Colonial Secretary's Office.[16] On taking and preparing specimens of lymph from tubercles on the left hand, wrist and ear lobe of Isabella Clarke, he concluded that "on microscopical examination I found abundant evidence of the presence of numerous leprosy bacilli densely packed within the tissue cells. These characteristic histological appearances prove conclusively that the patient is suffering from true Tubercular Leprosy."[17]

News of her diagnosis prompted local consternation and suspicion. Within the week, the Commissioner of Police and the Colonial Secretary received letters about her and her family from a member of the public. Mr Woods from the Criterion laundry, who asked that his name not be made public, found her very disturbing. To him, she was "no better than a common prostitute." He identified Bella's "boyfriend," and he also revealed that because she and her sister worked for the grocer, they may have handled vegetables.[18] As a woman, her sexuality was a source of contamination before her leprosy came into the picture; as a leper *and* a woman she represented a particularly potent source of contamination.

Bella Clarke was placed on Peel Island with a caretaker and his wife in July 1896. The Quarantine Station, the only other building on the island, was manned by an officer and his family. The doctor visited from Dunwich, on Stradbroke Island. The men with leprosy were held near Dunwich, which is on the adjacent island, but to all intents and purposes Bella was alone on Peel. But Bella was not someone who was going to disappear without attempting to be heard: in 1897–98 she besieged the Colonial Secretary with letters, and so did her mother. [19]

---

[16] "Letter from Government Bacteriologist C.J. Pound to the Col Sec," 30 September 1895 (*QSA* Col 272: No. 11665).

[17] "Report on Case of Leprosy – Isabella Clarke," *QSA* (Col 272: No. 11665).

[18] "Police report from Mr Woods, of the Criterion Laundry," *QSA* 5 October 1895 (Col 272: No. 10700 Commissioner of Police; No. 11976 Col Sec Office).

[19] *QSA*, "Letter," 3 March 1897 (Col 264: No. 03719); 25 March 1897 (Col 264: No. 04257); 14 September 1897 (Col 264: No. 11746); 20 September 1897 (Col 264: No. 12053); 14 March 1898 (Col 264: No. 05237); *QSA*, "Letters from Bella's mother," 11 November 1897 (Col 264: No. 14651); 5 February 1898. (Col 264: No. 01640); "Letter from Bella's mother to Dickson about her complaints about Drew and the caretakers," 4 April 1898 (Col 264).

Her story was brought to public attention by the *Street* newspaper. Initially, she was presented as a "girl" suffering the loneliness and isolation that the disease inevitably produced. In her first public appearance, the flavour of a tragic and romantic recluse surrounds her isolation on the island:

> Imagine yourself shut up on a lonely spot, precluded the society of your fellow beings, shunned by all, looked down upon as something loathsome, and fit only for death! Then, and then only – if her story be true – can you conceive the position of the lonely girl, shut out from society amidst the lazarette enclosures on Peel Island.[20]

Her published pathetic letters of appeal construct her as victim of an array of grotesque and insensitive captors: the heartless caretaker and his wife; the elderly residents of Dunwich who came over from Stradbroke Island to attend to her daily needs; an anonymous visitor to the island who "attempted to commit an indecent assault" upon her and made her the subject of his sexual boasts, and ultimately the colonial administration, represented by the uncaring doctor and the heartless Home Secretary.

In "the interests of the public," however, the newspaper soon moves away from the simple tale of injustice done to the poor, unfortunate leper girl to expose a story of political corruption in its investigation of the identity of the anonymous visitor who has attempted the indecent assault. At this point in the drama, there is a shift away from Bella, the wronged and suffering recluse, to George Drew, the son of a public servant, and the nepotistic beneficiary of a self-interested administration. In the process, however, Bella also undergoes a shift in signification. The earlier and more private medical construction of her sexuality and her illness converge so that she loses her identity and singularity and is subsumed into the threat that George Drew represents: both become figured as sources of contamination.

When the anonymous visitor was revealed by the intrepid newspaper to be the son of a senior Queensland Public Service Official, the newspaper mobilized a moral discourse which, in constructing itself as a fearless seeker of truth, offered both an exposé of immorality and a spirited defence of public health and the family in the face of political irresponsibility. It pointed out that, apart from the man's poor moral character, the danger that he represented to society by associating with Bella Clarke was "a danger to your families." Mere contact with him would be poisonous" (5). It appealed for

---

[20] "Queensland Girl Leper: Complains of Infamous Treatment: She Writes to the *Street*: We Want An Inquiry," *Street* (9 July 1898): 4.

an inquiry: "In the interests of the public, in the interests of our wives and children, our sisters, and brothers, we demand a public investigation into those terrible charges" (5).

The whole affair assumed the proportions of a public health scandal. The question of Drew's "connection" with Bella was the subject of the Colonial Secretary's inquiries of Doctor Smith, the medical officer at Dunwich and Peel. The possible contact of Mrs Walker, the Quarantine Officer's wife, with Bella was considered to be illegal (although there is nothing in the *Leprosy Act* on this matter). In addition, the newspaper reported meeting someone from the lazaret on the streets of the city, and it was horrified that they had been allowed the "priviledge of returning to public life."[21] Finally, there were rumours that men from the lazaret at Dunwich had actually visited Bella on Peel in their boat.[22]

When the Government appeared unresponsive to the newspaper's calls for an inquiry, an extremely well-attended public meeting was held and the outrage against the source of contamination that the woman on the island represented was cast as an exercise of citizenship and a protest against the policies of a Government that continued to permit the entry of indentured labour and Chinese into the colony. The circulation of the newspaper reportedly increased dramatically from 4,000, to 12,000, and then to 20,000, in a matter of days, and letters of support poured in from all over the colony.[23] The editorial imagined "A nation of tainted and corrupted beings, a mass of people – or rather creatures – whose fair features would be brutally disfigured by the evidence of loathsome disease." It roundly condemned

> a system of Government which – by the free importation of leper-infected aliens whose sole utilisation is for the starvation of clean white men to place dollars in well-filled pockets – not only winks at the advance of those disease-encircled hordes, but actually holds out its hand to welcome their presence here in Queensland.[24]

It accused the Government of

---

21 *Street* (30 July 1898): 5.

22 *QSA* Col 264. The 1898 Report of the *Lazarets at Dunwich and Peel Island* lists eleven men, including James Quigley and a thirteen-year-old child at Dunwich. In 1894, a *Report of Particulars Relating to Lepers Who Have Been Placed Under Restraint in the Colony* lists two at Dunwich.

23 *Street* (30 July 1898): 4.

24 "The Leprosy Scandal," *Street* (30 July 1898): 4.

> gradually contaminating you and your children and their children after them, and to the third, fourth, and aye, even more generations, with diseases that will render them satires on the image of God and repulsive lazars in the sight of men. A few years more and how many of you will be free from taint? Even as you sit cosy in your easy chair before a cheery fire, your children, endeared to you, may be contracting hideous diseases from association with others, who have intermingled and played about the pest-houses of human life in Queensland ...

Contagion and children were linked, for now the corruption of leprosy was indistinguishable from a frighteningly mixed racial type. At the meeting, people were asked if they were prepared to be complacent about this affront to their safety and their political identity: "to allow this fair colony to go forth in conjunction with such odious designations as leper land." As citizens, it was their responsibility to see that the laws on leprosy were "administered with the strictest observance."[25]

In the midst of this, when news of Bella's removal to the male lazaret at Dunwich was known, the Colonial Secretary and the Premier each received a letter from a member of the public requesting that she not be relocated at Dunwich for fear of the consequences: "I see in the newspapers that the female leper at Peel Island is to be sent to the Lazarette at Dunwich. There is a young man there in whom I take a very great interest and I do not like a young female to be sent near him. I need say no more[,] you will understand the rest. I was told sometime ago by an inmate of Dunwich that disgusting conduct is carried on there between the men and women. They meet somewhere in the scrub."[26]

His letter to the Premier disclosed more of his concern:

> There is young man there who is very dear to me and I do not like a young woman to be sent near him, his body is falling to pieces but I want his soul to be pure and holy.

He would not accept guarantees that what he feared would not occur:

> It is all nonsense to say that safeguards have been fully provided for the prevention of intercourse between the Lazarets for the different sexes. You cannot do it. Why not provide this poor girl with a place near the women's quarters at Dunwich? I would respectfully ask you not to allow this young woman to be sent to the men's Lazaret.[27]

---

25 "The Leprosy Atrocities," *Street* (20 August 1898): 4.

26 *QSA*, "Letter to the Home Sec from Denis Lineham," 1 June 1898 (Col 264: No. 07375).

27 *QSA*, 14 June 1898 (Col 264: No. 08021).

The response from the administration was that there was no need for alarm:

> I should advise the writer not to accept implicitly the reports he may hear circulated by dismissed or expelled patients and officers. Every care is taken at Dunwich of the moral as well as physical condition of the patients.
>
> Note: Inform Mr Lineham that after [...] inquiry the Premier has satisfied himself that the safeguards provided are amply sufficient [...] [to prevent] any possibility of intercourse between the sexes.[28]

Bella had become not only the source of leprotic contamination of the whole colony, but also the focus of ungovernable sexual temptation. She was, both morally and physically, a source of corruption – the heart of decay and degeneration lying in wait on Peel Island to be unleashed on the colony. The woman's sexuality by itself was sufficient to brand her a potential contaminant. Women, as Brinton notes, were also regarded as mothers of the race, and so Bella, though the contaminating source, might also be responsible for producing children who retained and, in turn, transmitted its impurity. Her leprosy, with its medieval sexual taint, in conjunction with her gender were combined into a powerful symbol of disease, degeneration and political corruption figured as female. At the same time, the *Worker* began to publish satirical cartoons of the Colonial Secretary (eventually to become the Premier of the colony) waltzing with a woman who represented both colony, leper, and prostitute (see Figures 1 and 2). Bella's body was the body politic corrupted; its damaged sexual and biological integrity became the integrity and heritage of the colony's body, prostituted for the sake of economics and the profit of a few powerful men.

The cartoons of the incident in the *Worker* show a disgusting, demonic figure with uplifted arms overshadowing, possessing and conjuring up the figures of the Premier, a kanaka and a Chinaman, all marked with the blotches of leprosy and indulging in a frenzied dance around the pedestal on which she stands. They are under her spell, the spell of disease, and she as woman, as prostitute and as the colony evokes the partnership between political influence and alien labour, which has prostituted and thereby jeopardized the heritage of the colony.

---

[28] Note: "The writer of this letter need be under no apprehension [...] safeguards have been fully prepared in connection therewith and indeed the woman will be under stricter surveillance than at Peel Island."

The sense of a community in peril from an inundation of "others" was also graphically depicted by the *Worker*. In "A More Disastrous Flood than That of '93," the homestead and family of the white settler are threatened by an endless sea of "coloured alien labour." The only protection for the imperilled home and its orderly domestic space is a fortress wall, which has already been breached by the growing flood.[29] Quarantine, as spiritual or moral stockade, and as protection from disease-bearing germs can be seen as the symptom of a fortress mentality which combined the sense of Australia as "pure" – a "clean slate" on which to build something that did not continue the problems of the past, a sort of social experiment – with a sense of the island continent as supporting a highly vulnerable, isolated society. Enclosed within the fortress, the colony found itself progressively relegating those elements of society outside its boundaries to the category of contaminants. The decayed space in the midst of the familiar and domestic exemplified by the carefully tended lawns and encircled by fruit tress, on the edges of which are the murky mangroves and the limitless, enduring ocean perfectly represents a nation's vision of itself.

So, in heritage and conservation terms, insignificant little Peel Island serves as more than evidence of a system of incarceration that belongs to the past. In *Green Imperialism*, Grove suggests that "an island is an analogue of society in which Paradise might be recreated or realised on earth."[30] In contrast, the history of Peel suggests that it is more an analogue of hell. In a hopelessly clumsy and belatedly rearguard action, the colonial authorities hysterically quarantined those who bore the marks of bodily degeneration in order to protect society from contaminating 'Others', while it nevertheless continued to capitalize on their labour. Peel Island is a microcosm of a whole island continent that still patrols its boundaries in an effort to maintain its racial composition, simultaneously sustaining its paranoia about the polluting effects of Others.

---

[29] "A More Disastrous Flood Than That of '93," *Worker* (26 November 1898): 4.

[30] R.H. Grove, *Green Imperialism: Colonial Expansion, Tropical Island Edens, and the Origins of Environmentalism, 1600–1860* (Cambridge & New York: Cambridge UP, 1995): 46.

WORKS CITED

Frank, Johann Peter. *A System of Complete Medical Police: Selections from Johann Peter Frank*, ed. Erna Lesky, tr. E. Vilim (1788; Baltimore MD: Johns Hopkins UP, 1970).

Grove, R.H. *Green Imperialism: Colonial Expansion, Tropical Island Edens, and the Origins of Environmentalism, 1600–1860* (Cambridge & New York: Cambridge UP, 1995).

"The Leprosy Atrocities," *Street* (20 August 1898): 4.

"Leprosy in New South Wales," *Australasian Medical Gazette* (June 1891): 280.

"The Leprosy Scandal," *Street* (30 July 1898): 4.

"A More Disastrous Flood Than That of '93," *Worker* (26 November 1898): 4.

Murray, D. Christie. *Daily Telegraph* (10 October 1891): 1.

"Presidential Address," *Australian Medical Journal* (15 February 1893): 74.

"Public Health in New South Wales," *Lancet* (10 October 1891): 827.

"Queensland Girl Leper: Complains of Infamous Treatment: She Writes to the *Street*: We Want An Inquiry," *Street* (9 July 1898): 4.

Queensland State Archives (QSA). Brisbane, Queensland.

Riddel, Robert. *'The Leper shall Dwell Alone': Peel Island Lazaret Conservation Plan: A Report for the Department of Environment and Heritage* (Brisbane, Queensland: Queensland Department of Environment and Heritage, 1993).

"The Animals Are Innocent"
Latter-Day Women Travellers in Africa*

GILLIAN WHITLOCK

AT THE VERY START of *Gorillas, Tea and Coffee*, Kate Llewellyn's travel diary from Africa, the Australian writer is "for the first time in my life" unable to speak. Arrival in Johannesburg and the minibus tour of Soweto leaves her "sitting in the back of the bus shrinking from what I had seen as a snail sprinkled with salt. I was practically hissing and fizzing."[1] Day three: on to Kenya, to Nairobi. Here traffic fumes, street children, and enquiries about her sex life from complete strangers produce more hissing and fizzing. Lonely Planet's *Africa on a Shoestring* warns: "at every border, at every turn the woman traveller is under suspicion."[2] In Nairobi two things bring Llewellyn relief, and bring her out of her shell to effect an entry into Africa. One is to find she can buy her favourite earplugs, the same French Quires she had bought in Paris, at a Nairobi chemist. The second discovery is the Karen Blixen Museum. Blixen's memoir *Out of Africa* is one of Llewellyn's favourite books, a twentieth-century masterpiece "so

* Note: This essay initiated the longer study printed as ch. 4 in Gillian Whitlock, *The Intimate Empire* (London: Cassell, 2000).

1 Kate Llewellyn, *Gorillas, Tea and Coffee: An African Sketchbook* (Melbourne: Hudson, 1996): 4.

2 Geoff Crowther, *Africa on a Shoestring* (Hawthorn, Victoria: Lonely Planet, 1989): 15.

full of love that it shines and shines." She finds the house stands "as if Karen Blixen had just walked out for a day." The white linen and lace spreads are ready in the guest room, the dining room stands "as if for guests, lace table cloth [...] roses embroidered on tapestry-covered straight-backed chairs. Danish blue and white china." Jodhpurs and pith helmet and tiger-skin rug hang in the bedroom; in the dining room a great lionskin is on the floor, "the head snarling or roaring with long white teeth."[3]

For Kuki Gallmann also, *Out of Africa* is embedded as a kind of template for her recent autobiography, *I Dreamed of Africa*. From the very outset, this recent narration of the migration from a degenerate post-industrial Europe to a Kenya which represents new possibilities for becoming "one's true self" draws on the tropes and figurations which are fundamental to Blixen's memoir. *Out of Africa* is there in Gallmann's dream of Africa not just in the epigraphs, or in what we are told in the cover blurb, but in a more extensive borrowing of trope and metaphor. Gallmann, too, will construct a narrative of belonging to Africa, of finding her true self there. She too will find a sensual expression of herself through liaison with a man who has "the ancient and refined wisdom of the warrior" – Paolo the Hunter – and through birthing a son who emerges as a natural *kijana*, a leader and superior consciousness. These contemporary expressions of the white hunter mythography are buried in the land in a ritual of possession and ownership which echoes the burial of Denys Finch Hatton in *Out of Africa*, "a land over which he shall reign for ever, and from which he shall never be separated."[4] Gallmann plants a yellow fever tree on each of these graves, "one day, its roots [...] would reach the body, which would nourish the tree and become part of the landscape."[5] This metaphor of burial as a means of claiming the land is a trope of settler writing in particular. In Gallmann's book it takes a distinctive turn in that these two trees, nourished by these white warriors, become the emblem of the Gallmann Memorial Foundation, which is dedicated to the conservation of threatened species in Kenya, and the "harmonious coexistence of man and the environment" (cover blurb). Gallmann's narrative culminates in "The Ivory Fire," the very public burning of twelve tons of elephant tusks in Nairobi in 1989 which was organized by Richard Leakey

[3] Karen Blixen [Isak Dinesen], *Out of Africa* (1937; Harmondsworth: Penguin, 1982): 19, 20.

[4] Kuki Gallmann, *I Dreamed of Africa* (Harmondsworth: Penguin, 1992): 179.

[5] Gallmann, *I Dreamed of Africa*, 129.

when he became Director of the Wildlife Department. For Gallmann, the blaze summarizes, purifies and makes sense of all that has happened in her life.

For both Kate Llewellyn and Kuki Gallmann *Out of Africa* is available as a vehicle for an imaginative entry into Africa. In imagining a part – Kenya – as representative for the entire continent, the quite uncharacteristic and specific nature of Kenya's recent and relatively short-lived existence as an elite settler colony in British East Africa is obscured. This colonization did not really get underway until early this century and was superseded by independence in 1963. Kenya, in particular the area known as the White Highlands, presents an environment which is ripe for European fantasy. For geographical and historical reasons, large populations of animals remained in Kenya long after they had been shot out or displaced in western and southern Africa. The natural environment there is particularly responsive to very different kinds of intellectual and physical uses of Africa for the West: on the one hand the plains and the highlands, the vegetation and the variety of animals have been available for romantic visions of the loss or discovery of an essential self in nature. On the other hand, they have been equally available to the requirements of sport hunting, where certain kinds of animals become game, and large tracts of land are opened by the safari. The intersections and collisions between these two appropriations of Kenyan space recur in white autobiographical writing. Both have been important to imagining ways that Western men and women – particularly a semi-aristocratic class fragment – might imagine they belong in Kenya. The idea of Kenya as the essence of Africa, spelt out by Blixen, has remained a powerful and recurrent metaphor.

Out of Africa resurfaces in a third and quite different recent writing scene, and one which undoes some of the suturing which is so well hidden in the original. In 1988 the remains of a young white tourist, Julie Ward, were discovered in Masai Mara game reserve. After travelling to Kenya with an overland safari, Ward had remained in Kenya to photograph the wild animals, and being left alone in the reserve she had disappeared. The story of the search for Ward, the discovery of her remains and the absurd claims by Kenyan officials that wild animals were responsible (against all forensic evidence but in a desperate attempt to preserve the image of game parks like Masai Mara as tourist havens) led to a long campaign for justice organized by Ward's father. This has been presented in his own account, *The Animals Are Innocent*, and in Jeremy Gavron's *Darkness in Eden*. Under these conditions, as we might expect, for Ward and Gavron Kenya emerges as the essence of a different kind of Africa: primordial; savage; only superficially civilized and rational. The Ward case causes the long history of ideas and associations which are released by the interaction of "lustful" black men and "defenceless" white women to be set loose again. In the Kenyan instance it opens a path from the 1990s back to representations of Mau Mau in the 1950s.[6]

It emerges, in the course of her father's enquiries, that Julie Ward was killed by park rangers not because of their uncontrollable desire for a white woman, but because she probably witnessed their dealing in ivory. Hence the association in John Ward's book between the innocence of the animals and the white woman, over and against the preparedness of black men to slaughter both. The penultimate scene of Gavron's *Darkness in Eden* is Leakey's ivory fire, which he reads as a sign that Kenya's parks and reserves were about to become safe havens for elephants and tourists.

Under these conditions, where things fall apart, Blixen's fantasy of belonging in Africa as an independent woman and her presentation of Kenya as an Edenic place are juxtaposed quite explicitly with that other narrative, supposedly the reality to her romance, which presents the horror of Africa, figured in the white woman as subject to predatory black men. Rather than separating these in terms of romance and realism, we should perhaps begin to constitute both in terms of the powerful social imageries that Kenyan colonists used to construct narratives of domination and subordination in the borderlands between European and African knowledge and experience. Blixen's achievement is the construction of a memoir which makes the conditions of its being not only natural but noble.

That *Out of Africa*, rather than *Africa on a Shoestring*, should effect Kate Llewellyn's entry into African space should not surprise us. The former presents Africa as an empowering space for women. The latter foregrounds the threats to white women experienced in these borderlands. Since its first publication in 1937, Blixen's memoir stands as the most powerful and seductive capture of Kenya and, arguably, the most influential invention of African space for Western readers: "Africa distilled up through six thousand feet,

[6] Around Mau Mau, too, relations between black and white were represented in figures of the white woman and the black rapist. Fears of the thin veneer of civilization in Kenya were accentuated following the murder of a white woman and her family in their lonely farmhouse in 1952. In 1992, as in 1952, the discovery of a lock of blond hair from the victim is axiomatic, this piece of evidence becomes metonymic of the femininity, whiteness and innocence of the victim and the violence and depravity of black men.

like the strong and refined essence of a continent."[7] Llewellyn's recollection of the museum display of Blixen's house draws out the components of this seduction, and why the house might initiate the entry into African space which she desires. Readers of *Out of Africa* know the dining-room well: it is where Kamante, the Kikuyu houseboy Blixen transformed into a royal cook, served *omelette à la chasseur*. The china and linen were part of what Berkeley Cole called his sylvan retreat, where he and Finch Hatton could listen to the latest gramophone records from Europe and where Blixen, against all probability, induced peonies to grow (but not propagate) in the African garden. What is captured in the museum is the aristocratic plenitude which is depicted so graphically in the book, and it needs to be associated with the supremacist vision attached to that domestic setting. The author tells us that as she and Finch Hatton linger at the table their talk is so incandescent that they imagine "the wild Masai tribe, in their manyatta under the hills, would see the house all afire, like a star in the night, as the peasants of Umbria saw the house wherein Saint Francis and Saint Clare were entertaining one another upon theology."

As we have seen, the incandescence and the illusion of Blixen's presence remain not with the Masai but her fellow non-indigenous travellers. This, despite the fact that the house is progressively stripped bare in the final stories in *Out of Africa*, hollowed out like a skull, and the grass grows up to the doorstep. In dismantling the place she occupied, she also dismembers herself – that precarious identity which belonged in Africa for a relatively short time: "I myself was the lightest thing of all to get rid of."[8] Like the peonies, the vision of establishing a viable coffee plantation and a semi-aristocratic retreat in the Ngong Hills would be a brief flowering which would disappear completely. The viability and sustainability of the elitist version of settler colonialism which was planned for Kenya was always problematical. *Out of Africa* presents not only one of the most stunning and seductive visions of how a European woman might declare "This is where I belong," "This is where I ought to be," but also the failure and dissolution of that fantasy, and her return to Europe.

The case has been well made that Blixen was involved in a colonialist project which she simultaneously participated in, benefited from, despised, and

---

[7] Blixen, *Out of Africa*, 13.

[8] *Out of Africa*, 322.

repeatedly sought to subvert.[9] What Llewellyn views at the museum are the trappings of an elite settler society, a demographic fragment which relished the opportunity to turn away from industrialized metropolitan Europe and establish a pre-industrial, quasi-feudal order. This is a fantasy world, which *Out of Africa* not only constructs but also dismantles.

That Llewellyn fails to connect these artefacts to a classed, gendered and racial organization of settler culture interconnects with arguments I have made in other work about the uses of Africa in Australian women's travel writing, and with the strategic forgetting produced by the mirage which Blixen's tapestry animals inhabit. Beset as she is by a struggle to enter Africa imaginatively, to become a writer there, what Llewellyn responds to is a place for herself as that absent guest. The display is designed to suggest that the house awaits occupation still, that some may enter Blixen's magic circle and find their place at that table, in that bedroom where the jodhpurs and pith helmet and tigerskin rug hang.

In this, the display is a synecdoche of the memoir. Few texts signal a place for the narratee to enter and listen as brilliantly as *Out of Africa*. Ross Chambers points out that the ongoing readability of texts is crucially dependent upon recruiting the narratee; it is upon this that the ongoing authority of the narrator depends.[10] The autobiographical narrator of *Out of Africa* is a brilliant storyteller, her house is not only a repository for books but also a place designed for the telling and hearing of stories. Recent re-readings of *Out of Africa* as an oppositional text draw on feminist and psychoanalytic literary theories to authorize this narrator anew. From these perspectives, Dinesen's stories and memoirs are read as more complex, self-reflexive texts, anticipating not only feminist concerns with women's experience in androcentric societies, but also conceptions of language and subjectivity associated with the writings of Cixous, Kristeva and Irigaray.[11] This re-birthing of Blixen's texts as oppositional narratives authorizes her quite differently and brings a new recruitment of narratees to the feast. However, they must jostle aside an intradiegetic narratee for whom the place was set, the object of seduction, Denys Finch Hatton. They must come to terms with the relation-

---

[9] Susan Hardy Aiken, *Isak Dinesen and the Engendering of Narrative* (Chicago: U of Chicago P, 1990): 213.

[10] Ross Chambers, *Room for Maneuver: Reading Oppositional Narrative* (Chicago: U of Chicago P, 1991): 11.

[11] Aiken, *Isak Dinesen and the Engendering of Narrative*, 8.

ship between the fine china and linen on the table and the great lionskin beneath their feet.

Familiar as we are with the figure of the white hunter in the masculinist imperial creations of Rider Haggard and Ernest Hemingway, we miss the significance of his appearance in autobiographical writing by Blixen and her contemporary, Beryl Markham, and in Kuki Gallmann's more recent memoir. However, the desire to please the white hunter as an intradiegetic narratee is fundamental to the writing of *Out of Africa,* which sets before us an extravagant and diverse feast of stories, fables and characters. The logic of the memoir and, I suggest, its ongoing power to seduce and capture readers, is produced by the shape of this writing scene and the profoundly sensual and mythic invention of African space which it produces to satisfy the aristocratic taste of Denys, a listener of discrimination, who tutors the narrator in Latin, the Bible and the Greek poets, and makes sure the farm is stocked with the best in wine and tobacco. On his return, the farm becomes a metonym for the narrator: "it gave out what was in it; it spoke. [...] When I was expecting Denys back, and heard his car coming up the drive, I heard at the same time, the things of the farm all telling what they really were."[12]

This brings me back to the mirrors, to the work of figuration in the text. I have argued elsewhere that colonial encounters foreground processes of scripting the self through connections and unexpected interdependencies. For both Blixen and Beryl Markham, the production of the self in the autobiographical text is crucially dependent upon their figuration of Finch Hatton. The invention of Finch Hatton as the archetype of the hunter is dependent not upon his own writings at all, nor those passing references to him as the guide of choice for the elite hunting parties which set out from Nairobi. Others who made his acquaintance find these textual magnifications of him idiosyncratic. He is nevertheless a vital component in the way that both Blixen and Markham authorize themselves. The incorporation of Finch Hatton into *Out of Africa* as the narratee makes the nature of the exchange very clear. A kind of symbolic economy is at work in the text between narrator and narratee. The hunter brings to the farm the trophies of the hunt: "leopard and cheetah skins, to be made into fur coats in Paris, snake and lizard skins for shoes, and marabout feathers"[13] and in return Blixen experiments with many curious recipes out of old cookery books, and

12 Blixen, *Out of Africa*, 194.

13 *Out of Africa*, 179.

works to make European flowers grow in the garden. For Finch Hatton, the exchange includes stories, and the transformation of the narrator into Scheherazade, the woman who tells stories for her life, for their mutual pleasure. It should not surprise us that the death of the narratee produces the symbolic death of the narrator. As the house is emptied and the ambience dissipates, the narrator herself is dismembered, expelled from the nobility of Africa to the degradations of Europe, just like the giraffes in the little sketch "The Giraffes go to Hamburg." The white hunter, on the other hand, attains immortality and indigeneity through one of the tropes of settler writing: the burial where he becomes part of Africa and leaves his own mark upon it. His place as a natural aristocrat is confirmed when the lions habitually come to the grave "and make him an African monument."[14] He also, of course, has a small memorial at Eton.

The transmutation of the relationship between Blixen and Finch Hatton into a liaison between two quite different ways of occupying Kenyan space in that period of early settlement, *and* into the textual apparatus of narrator and narratee, is a brilliant innovation which is fundamental to the work of smoke and mirrors in *Out of Africa.* By smoke and mirrors I mean the processes of erasure and figuration which occur to naturalize the colonial relation which is being established here, which produce the effects of truth, authenticity and inevitability.

Blixen is quite specific that the safari and the farm are two mutually exclusive domains. Her own passion for safari and the kill is superseded by the domestic relation to space and place on the farm. We see little, then, of the quite different economy and relation between colonization and environment of which Finch Hatton was a part: the safari. Even as the establishment of settler colonization based on small-scale farms was proving difficult to establish, this quite different and distinctively modern management of land and animals was underway in Kenya. Hatton was a big game hunter, pioneering the use of the airplane to spot game. Tourism, in particular catering for the European elite in search of game, would be the industry which would ensure the ongoing viability of white settlement in Kenya.[15] This is, as it turns out, where they would "belong."

---

[14] Blixen, *Out of Africa*, 308.

[15] It has been estimated that 25 percent of the white population in Kenya in 1990 was dependent on the tourist industry.

West with the Night (1992), Beryl Markham's memoir, is no rival to *Out of Africa*'s status as a classic text. It does, however, open up a slightly different angle on how a new economy was emerging around the white hunter in this period. Markham was an important part of Finch Hatton's project to use the latest technology to deliver what his clients most desired: tusks. His intention was to modernize the management of the hunt, to spot elephants "efficiently" from the air and to offer his clients a new service: keeping a hunting party in touch with a moving herd. Ironically, Finch Hatton's death means that it is Blix, Baron von Blixen, who establishes a new frontier between hunting and tourism in Kenya, and a quite different relationship between the human and the animal in the hunt. The practicalities are clear: in Markham's words, "As a White Hunter, his job was to produce the game desired and to point it out to his employer of the moment."[16] There is nothing noble about this. The size and pace of the elephant made it available to a kind of commodification which is distinctively modern. There is no place for this in *Out of Africa.* The skins which Llewellyn sees on the floor and on the wall beside the pith helmet in the reconstruction of Blixen's house are the trophies of the elite hunt; the orgasmic pleasure of the kill for Blixen is produced by the purer, more aristocratic hunt for lions, which becomes a carefully choreographed, elaborate metaphor of intercourse and metamorphosis between the narrator and narratee.

However, Blixen's fantasy of belonging in Africa, of establishing a house which is "one with the landscape, so that nobody could tell where the one stopped and the other began,"[17] lends itself to a quite different emergent sensibility, which establishes a different kind of relationship between the human and animal worlds. The associations between romanticism and conservation go back to the nineteenth century, well before Blixen. However, the supremely self-centred insertion of the self into Africa which is performed in the memoir leads to a highly aestheticized presentation of the natural world. Animals are vital to the Edenic aestheticized style of the book. Giraffe are long-stemmed gigantic flowers. Rhinos are like big angular stones rollicking in the long valley. Leopards sitting on the roads are like tapestry animals. The fawn has a nose like a truffle. Gazelles are like toy animals on a billiard table. The nearness of the animals, the illusion of oneness with the natural world, represents a kind of intellectual and emotional use of Kenyan

---

16 Beryl Markham, *West with the Night* (London: Virago, 1992): 213.

17 Blixen, *Out of Africa*, 74.

landscapes and leads to a different type of commodification of African space. Blixen captures the animal world for the pleasure of the European gaze – in *Out of Africa* the animals seen on safari become available sights, their proximity ennobling: "the big game was out there still, in their own country; I could go and look them up once more if I liked. Their nearness gave a shine and play to the atmosphere of the farm."[18] Blixen is one of the first to translate the new technology of the camera into a modern version of the hunt: "The civilized people have lost the aptitude of stillness, and must take lessons in silence from the wild before they are accepted by it. The art of moving gently, without suddenness, is the first to be studied by the hunter, and more so by the hunter with the camera" (24). In this way, Blixen is a harbinger of tastes and sensibilities which we recognize still, and which produces the Kenya which would be so attractive for modern tourism and so seductive to the likes of Julie Ward.

It is tempting, of course, to see the white hunter as a masculinist anachronism, superseded by a modern, feminized consciousness where the only trophy of the hunt is a photograph. The lesson of *Out of Africa* is that the invention of Kenya in colonial discourse has been a complex intersection of very different fantasies, and the opposition between masculine and feminine, past and present and – I would suggest – exploitative and enlightened inventions of Kenya are called into question. Rather, they operate as coordinates within the same discursive web, one which remains deeply implicated in the ongoing fantasy of white supremacy in black Africa, and the interests of that small settler fragment. In this world, as Kuki Gallmann reminds us, a passion for hunting "might logically evolve into pure conservationism."[19] Both Gallmann and Blixen literally and figuratively bury the white hunter and ensure that white writing about Kenya will continue to feed on his bones. Gavron's narrative of the Ward murder, which radically calls into question fantasies about white women in particular "belonging" in Kenya, operates within the same discursive frame through the elision of wildlife and the tourist – the shared rights of both elephants and Julie Ward to roam free as innocents in the Masai Mara.

These are, of course, the very tensions which sent Kate Llewellyn to the Karen Blixen museum in the first place: the questions about her sex life from strangers on the streets of Nairobi; the confirmation from the Lonely

---

[18] Blixen, *Out of Africa*, 23.

[19] Gallmann, *I Dreamed of Africa*, 56.

Planet guide that as a lone woman traveller she is in a precarious situation. Defeated by the "complexity, the vastness, the entire universal rave that is Africa," Llewellyn returns home to her kitchen in Leura to cut up tomatoes and think about it, "the smell of Gay Bilson's adaptation of Philip Searle's roast tomato chutney fills the house."[20] When I read bits of Llewellyn to my husband he said, "why dear, she's just like you." So she is. In practice. But not in theory. There I am with Dervla Murphy, who gets off the plane in Nairobi with her sixtieth-birthday present, a Dawes Ascent mountain bike, the cyclist's equivalent of a Rolls–Royce, named Lear. Lear and Murphy make their way to a Christian guesthouse on Bishop's Road, where Lear is locked to the bed and Murphy spends the day ambling around Nairobi, getting beaten by paramilitaries on the way. That night she watches television:

> Julie Ward's father was being interviewed outside Nairobi's imposing courthouse. On that flickering screen the poor fellow seemed to have St Vitus's Dance. The Luo teacher beside me muttered angrily, "Why so much fuss about this one murder? Only because she was White and killing her is bad for the tourist trade. Everyday Kenyan people are murdered in Kenya and there is no fuss!"[21]

Murphy comes to Africa from Northern Ireland; she recognizes smoke and mirrors, and cycles out of Nairobi the next day without visiting the Karen Blixen museum. As she later points out, the settlers' view of the African, absorbed from Blixen, renewed by Kuki Gallmann, focuses on the land and its animals and remains well within the comfort zone of white readers. "This endangers the Africans who, as I write these words, are being indirectly killed, in considerable numbers, by relentless First World profiteering."[22] The animals may be innocent, but we are not.

---

[20] Llewellyn, *Gorillas, Tea and Coffee*, 215.

[21] Dervla Murphy, *The Ukimwi Road: From Kenya to Zimbabwe* (London: Flamingo, 1994): 6.

[22] Murphy, *The Ukimwi Road*, 38.

WORKS CITED

Aiken, Susan Hardy. *Isak Dinesen and the Engendering of Narrative* (Chicago: U of Chicago P, 1990).

Blixen, Karen [Isak Dinesen]. *Out of Africa* (1937; Harmondsworth: Penguin, 1982).

Chambers, Ross. *Room for Maneuver: Reading Oppositional Narrative* (Chicago: U of Chicago P, 1991).

Crowther, Geoff. *Africa on a Shoestring* (Hawthorn, Victoria: Lonely Planet, 1989).

Gallmann, Kuki. *I Dreamed of Africa* (Harmondsworth: Penguin, 1992).

Gavron, Jeremy. *Darkness in Eden: The Murder of Julie Ward* (London: HarperCollins, 1991).

Huxley, Elspeth. *The Flame Trees of Thika: Memories of an African Childhood* (Harmondsworth: Penguin, 1962).

Kennedy, Dane. *Islands of White: Settler Society and Culture in Kenya and Southern Rhodesia, 1890–1939* (Durham NC: Duke UP, 1987).

Llewellyn, Kate. *Gorillas, Tea and Coffee: An African Sketchbook* (Melbourne: Hudson, 1996).

Markham, Beryl. *West With The Night* (London: Virago, 1992).

Murphy, Dervla. *The Ukimwi Road: From Kenya to Zimbabwe* (London: Flamingo, 1994).

Shaw, Carolyn Martin. *Colonial Inscriptions: Race, Sex and Class in Kenya* (Minneapolis: U of Minnesota P, 1995).

Smith, Sidonie, "The Other Woman and the Racial Politics of Gender: Isak Dinesen and Beryl Markham in Kenya," in *De/Colonizing the Subject: The Politics of Gender in Women's Autobiography*, ed. Sidonie Smith & Julia Watson (Minneapolis: U of Minnesota P, 1992): 410–35.

Tidrick, Kathryn. *Empire and the English Character* (London: I.B. Tauris, 1992).

Ward, John. *The Animals Are Innocent: The Search for Julie's Killers* (London: BCA, 1991).

Contributors

RUTH BLAIR teaches American and environmental literatures at the University of Queensland. She has written on Herman Melville in the Pacific and edited the Oxford World's Classics edition of Melville's *Typee*. She is working on an anthology of Australian nature writing.

CLAUDIA BRANDENSTEIN is completing a doctorate in the School of English, Media Studies, and Art History at the University of Queensland in nineteenth-century travel writing about the Caribbean. She has published on Charles Kingsley and Lady Maria Nugent.

LEIGH DALE teaches literature, and is Program Director for Contemporary Studies, in the School of English, Media Studies and Art History at the University of Queensland in Brisbane, and is the editor of the journal *Australian Literary Studies*. Publications include *The Body in the Library* (1998; co-edited with Simon Ryan) and *Re-Cognition: New Essays in Australian Studies* (2005; edited with Margaret Henderson).

CARRIE DAWSON is Associate Professor for English and Canadian Studies at Dalhousie University. Her teaching and research areas include Canadian literatures, contemporary fiction, postcolonial theory, and ecocriticism. She is the editor, with Maggie Nolan, of *Who's Who? Hoaxes, Imposture and Identity Crises in Australian Literature* (2004), and has published articles on Alice Munro, Michael Ondaatje, Grey Owl, and (with Marjorie Stone) Canadian multiculturalism.

ROBERT DIXON is Australian Research Council Professorial Fellow in the School of English, Media Studies & Art History and the Australian Studies Centre at the University of Queensland. His most recent book is *Prosthetic Gods: Travel, Representation and Colonial Governance* (2001). He is currently writing a book about Frank Hurley, *Travelling Mass-Media Circus*, and preparing an illustrated edition of Hurley's diaries.

HELEN GILBERT is Lecturer in Drama and Theatre Studies, University of Queensland. Her research interests include drama and performance, contemporary drama (especially Australian, Canadian, Caribbean), colonial nursing, and postcolonial theory. She is the author of *Sightlines: Race, Gender and Nation in Contemporary Australian Theatre* (1998) and, with Joanne Tompkins, of *Post-Colonial Drama: Theory, Practice, Politics* (1996). She has also edited *Post-Colonial Stages: Critical & Creative Views on Drama, Theatre and Performance* (1999), *Postcolonial Plays: An Anthology* (2001), *In Transit: Travel, Text, Empire* (2002; with Anna Johnston), and *Diaspora: Negotiating Asian-Australia* (2000; with Tseen Khoo & Jacqueline Lo).

CATHERINE HOWELL is a South Australian, with a Cambridge degree in French literature and postcolonial studies, who has lived and worked in the UK since 2000, currently as an educational technologist and project liaison officer at the University of Cambridge.

ANNA JOHNSTON is Senior Lecturer in the School of English, Journalism and European Languages at the University of Tasmania. Her research interests are in colonial and postcolonial literatures and cultures. She is a member of the Executive of the International Australian Studies Association and Deputy Director of the Centre for Colonialism and its Aftermath (UTAS). She is co-editor (with Helen Gilbert) of *In Transit: Travel, Text, Empire* (2002), and is currently co-editor (with Ralph Crane) of *new literatures review*.

ANDREW MCCANN is Associate Professor in the Department of English at Dartmouth College. He is author of *Cultural Politics in the 1790s: Literature, Radicalism and the Public Sphere* (1999), *Marcus Clarke's Bohemia: Literature and Modernity in Colonial Melbourne* (2004) and two novels, *The White Body of Evening* (2002) and *Subtopia* (2005).

SUSI O'BRIEN is an Assistant Professor in the Department of English and Cultural Studies at McMaster University. Her research interests include postcolonialism, ecocriticism, and cultural studies, with a particular focus on the relationship between these areas and processes of globalization and the representation of animals and the environment in postcolonial fiction. She is currently completing a book, titled "Imagined Diversities: Ecology, Postcolonialism and the Production of Global Culture," which looks at the incorporation of postcolonialist and ecological rhetoric into narratives (literary, popular and academic) about the idea of global citizenship. With Imre Szeman she has written *A User's Guide to Popular Culture* (2003).

JO ROBERTSON is the coordinator of the International Leprosy Association's Global Project on the History of Leprosy. Her most recent publications are "Culion, the 'Island of the Living Dead': or Another Look at

Leprosy and Citizenship," in *Politics of Prevention, Health Propaganda and the Organisation of Hospitals 1800–2000*, ed. Astri Andresen et al. (2005), and, with A. Colin McDougall, "Leprosy Work and Research in Oxford, The United Kingdom: Four Decades in the Pursuit of New Knowledge about an Arcane Disease," in the *International Journal of Dermatology* (2005).

MEENAKSHI SHARMA, a graduate of the University of Queensland, is currently Associate Professor at the Institute of Management Technology, Ghaziabad. Her research and publications have been in the areas of postcolonial theory and Indian writing in English, including the Indian diaspora and the representation of tradition and modernity. She is the editor of a volume of interviews with contemporary Indian authors, *The Wordsmiths* (1996), and the author of *Postcolonial Indian Writing: Between Co-Option and Resistance* (2003).

CHRIS TIFFIN is Director of Postgraduate Studies in the Department of English at the University of Queensland, Australia. He is the author of articles on colonial and postcolonial literatures and on bibliography, has compiled the standard bibliography on Rosa Praed (2005), and has edited *South Pacific Images* (1978). He is also the co-editor of *South Pacific Stories* (1980; with Helen Tiffin) and of *De-Scribing Empire: Post-Colonialism and Textuality* (1994; with Alan Lawson). Research interests include theory of editing, humanities and computing, nineteenth century British literature (especially publishing history), and Australian literature; current research includes nineteenth-century colonialism, the design of electronic scholarly editions, and electronic textuality.

HELEN TIFFIN, more recently at Queen's University, Kingston, Ontario, now teaches at the University of Tasmania. She is the co-author, with Bill Ashcroft and Gareth Griffiths, of *The Empire Writes Back: Post-Colonial Literatures, Theory & Practice* (1989; rev. 2002), *Key Concepts in Post-Colonial Studies* (1998), *Post-Colonial Literatures in English: General Theoretical & Comparative 1970–1993* (1997), and, with Diana Brydon, of *Decolonising Fictions: Comparative Studies in Post-Colonial Literatures* (1993). Her co-edited volumes include *Past the Last Post: Theorizing Post-Colonialism and Post-Modernism* (1991; with Ian Adam), *The Post-Colonial Studies Reader* (1995, rev. 2005; with Bill Ashcroft and Gareth Griffiths), *Re-Siting Queen's English: Text and Tradition in Post-Colonial Literatures* (1992; with Gillian Whitlock), *South Pacific Stories* (1980; with Chris Tiffin), and *After Europe: Critical Theory and Post-Colonial Writing* (1989; with Stephen Slemon and Diana Brydon). Helen Tiffin is currently working on a book on the postcolonial novel. Her research interests include postcolonial literatures and literary theory, Caribbean studies, disease in colonialist contexts, representation of animals, and the species boundary.

GILLIAN WHITLOCK is Professor in the School of English, Media Studies and Art History at the University of Queensland. Her research interests include autobiography and biography, postcolonial studies, and women's writing. She is the author of *The Intimate Empire: Reading Women's Autobiography* (2000), and co-editor of *Australian/Canadian Literatures in English: Comparative Perspectives* (1987; with Russell McDougall), *Re-Siting Queen's English: Text and Tradition in Post-Colonial Literatures* (1992; with Helen Tiffin), *Images of Australia* (1992; with David Carter), and *Uncertain Beginnings: Debates in Australian Studies* (1993; with Gail Reekie). She is editor of *Autographs: Contemporary Australian Autobiography* (1996, rev. 2000) and *Eight Voice of the Eighties: Contemporary Australian Women's Writing* (1989).

Index